spineless
wonders

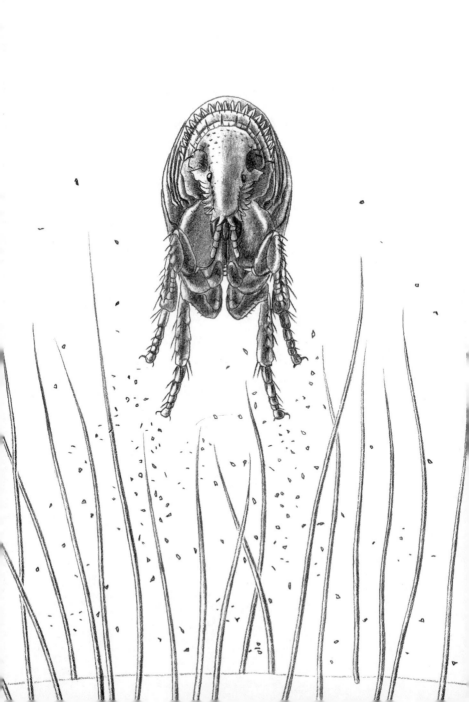

spineless
wonders

Strange Tales from the
Invertebrate World

Richard Conniff

Illustrations by Sally Bensusen

Henry Holt and Company   New York

Henry Holt and Company, Inc.
Publishers since 1866
115 West 18th Street
New York, New York 10011

Henry Holt® is a registered trademark of
Henry Holt and Company, Inc.

Published in Canada by Fitzhenry & Whiteside Ltd.,
195 Allstate Parkway, Markham, Ontario L3R 4T8.

Library of Congress Cataloging-in-Publication Data
Conniff, Richard.
Spineless wonders: strange tales from the invertebrate world /
Richard Conniff; illustrations by Sally Bensusen.—1st ed.
p.   cm.
Includes bibliographical references and index.
1. Invertebrates.   I. Title.
QL362.C66   1996   96-11748
592—dc20   CIP

ISBN 0-8050-4218-0

Henry Holt books are available for special
promotions and premiums. For details contact:
Director, Special Markets.

First Edition—1996

Designed by Paula R. Szafranski

Printed in the United States of America
All first editions are printed on acid-free paper. ∞

3   5   7   9   10   8   6   4

contents

acknowledgments

I could not have completed the research that went into this book without considerable outside help. At *Smithsonian* magazine, I am deeply grateful to Don Moser, the editor who assigned me to write about moths, with the caveat, "just don't make me look at them," and who sometimes wondered if I couldn't choose a subject "bigger than a breadbox," but nonetheless sent me off to report on fire ants and fleas. Also at *Smithsonian,* Jim Doherty went well beyond the call of duty to keep me steadily working, decently paid, and judiciously insulted. Doherty is that rare thing among editors, a vertebrate. He was ably assisted by Sally Maran, Beth Py-Lieberman, Marian Holmes, Fran Glennon, Ed Rich, Bonnie Stutski, and Marika Carley.

At *Audubon* magazine, Les Line and Gary Soucie were an enthusiastic audience for some of my more repellent subject matter. At *National Geographic* magazine, Bob Poole assigned me to write about tarantulas and has struggled nobly for the cause of good writing. This is also the third magazine where I have worked with Jennifer Reek, who describes herself as my "intellectual domestic." Mel Allen at *Yankee* magazine generously supported my work, as did Gil Rogin and Peter Carey at *Discover* magazine, and Steve Burns and Mike Quattrone at the Discovery Channel. At National Geographic Television, past and present, I have received considerable help from Lis Leader, Stella Cha, Chris Weber, Keenan Smart, Kevin Krug, Tom Simon, Michael Rosenfeld, and Tim Kelly. I also want to thank my agent, John Thornton, of the Spieler Agency and my editor, Ray Roberts, of Henry Holt and Co., who together got this book off the ground, and Sally Bensusen, whose illustrations managed to be accurate, lively, and also often lovely (the subject matter notwithstanding). Jane Maclellan, at the time an undergraduate at Connecticut College, helped pull together the loose ends of the manuscript. The librarians of Deep River and Essex, Connecticut, particularly Sarah Adams, Susie Smith, and Mary Attridge, routinely fielded my research questions.

For their generous help on specific stories, I am most deeply obliged to the scientists who did the basic research and took the time to explain it to me. On *moths:* David Wagner of the University of Connecticut, Doug Ferguson and Don Davis of the National Museum of Natural History, Tom Eisner of Cornell University, James Fullard of the University of Toronto, and Noel McFarland. *Houseflies:* Bernard Greenberg of the University of Illinois at Chicago, Vincent Dethier of the University of Massachusetts. *Leeches:* Joseph Upton, M.D., of Children's Hospital in Boston, Roy Sawyer of Biopharm, Gunther Stent at the University of California at Berkeley, Bill Kristan at the University of California at San Diego, and Donald Klemm at the U.S. Environmental Protection Agency. *Worms:* Sam James of Maharishi International University and Alan Tomlin of Agriculture and Agri-Food Canada. *Beetles:* Terry Erwin and George Venable of the National Museum of Natural History. *Fleas:* Michael Rust of the University of California at Riverside, Miriam Rothschild, and Charles Neville of the University of Bristol. *Dragonflies:* Mike May of Rutgers University, and Ken Soltesz, Jerrell Daigle, Carl Schiffer, Steve Valley, John Michalski, and Dennis Paulson of the Dragonfly Society of America. *Tarantulas:* Rick West of the Royal British Columbia Museum, Bill Lamar of the University of Texas at Tyler, Mark Moffet of the Museum of Comparative Zoology at Harvard University, and Norman Platnick of the American Museum of Natural History. *Mosquitoes:* Wayne Crans of Rutgers University. *Slime eels:* John B. Heiser and Ric Martini of Cornell University's Shoals Marine Laboratory and Susumu Kato of the National Marine Fisheries Service. *Fire ants:* Walter Tschinkel and Debbie Cassill of Florida State University, and Sanford Porter, David Williams, and Bob Vander Meer of the U.S. Department of Agriculture. *Giant squid:* Clyde Roper of the National Museum of Natural History and Mary Rice of the Smithsonian Marine Station in Fort Pierce, Florida.

Finally, I owe thanks and much more to my wife, Karen, and our children for tolerating my frequent absences, eccentric work habits, and unsuitable invertebrate house guests.

spineless
wonders

introduction

The Joy of Formication

It was late on a Friday afternoon not long ago in the American Southwest. I was out reporting on behalf of a well-known American magazine, together with a crew from an equally well-known British television company. We had spent the past five days attempting to observe large, hostile wasps stinging and parasitizing large, hairy tarantulas.

We had tried to get them to do it indoors and outdoors, with big animals and small, with artificial burrows and with real ones, down which our favorite wasp disappeared and never returned, as if it had struck up a friendship with the tarantula, or a small card game, in conspiracy against the ludicrous human beings overhead. (It is of course also possible that the tarantula had turned the tables and eaten the wasp.)

In the ninety-degree heat, under the cloudless sky and unfiltered desert sun, wasp and tarantula had reduced us by Friday afternoon to a state of gibbering hilarity. We placed a tarantula down an artificial burrow out amid the saguaro and the scrubby creosote bush. Our wasp waited impatiently in an upside-down two-gallon plastic jar with the lid off. To calm the wasp, we draped a dark cloth over the jar and the adjacent tarantula hole.

Then we all got into position. The producer put up his green, blue, and white golf umbrella to shade the scene of battle. The cameraman knelt down, and the assistant producer advised him that if he backed up a foot, he was liable to be molested by a cholla cactus, possibly losing focus.

When everything was ready, the producer whisked the cloth covering off the jar, like a magician removing a tablecloth. Then he moved the jar and the wasp directly over the tarantula hole. All of us held our breaths. The wasp went down the hole. But no tarantula came surging out to defend itself. "Let's give it a minute," someone said, and we waited with reaffirmed purposefulness. Nothing happened.

Finally, the assistant producer looked up and said, "Why is there a tarantula hanging in that bush over there?" All of us lifted our heads and turned. It was our tarantula, flicked away on the cloth in a daring attempt at aerial escape. The spider gods were against us.

It was at this precise moment, as I knelt in the middle of the desert, at the tail end of a miserably hot week spent not merely studying invertebrates but also being made a fool by them, that it dawned on me what had become of my life. I let out a sob.

Or wait . . .

Maybe it wasn't that day at all. Maybe it was the moment in the Border Country of northern England, when I found myself running across a farm where the sheep were spread out like gravestones on the misty hillside, cradling in my hands a wriggling, cold lump of 133 earthworms.

Or it may have been the time in Peru when my host served us palm beetle grubs, presented first for our inspection in their natural state, finger-long and with a yellow, puslike complexion.

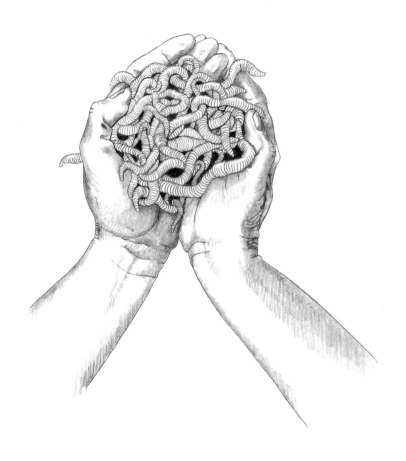

Lightly sautéed, the shriveled skin was nicely al dente, and the innards had the consistency of warm mayonnaise. To be polite, I took two, and wondered once again how I had come to this juncture.

I grew up in the suburbs of New Jersey, where the background hum to which my inner being resonated on summer afternoons wasn't crickets but the sound of trucks and cars pouring down Route 3 into Manhattan. I was never the kind of child who chases butterflies, or feeds tadpoles to dragonfly nymphs. The natural world was absent from my upbringing, both physically and philosophically. My father liked to quote

Horace: "You can pitch Nature out with a fork, but she will always come back," and, by God, we pitched.

I suspect, regrettably, that the absence of nature was typical for my generation of Americans, born in the suburbanized decade or so after World War II. Animals even in the agricultural context were alien. In college, I studied politics and English poetry. We called ourselves environmentalists, but zoology, and above all *invertebrate* zoology, were subjects for geeks.

So I am a latecomer to the field. My education began when I walked away from my job at a daily newspaper, where the diet of uninspired political corruption and routine murders had worn thin. Writing about the natural world was a form of relief at first. Maybe it shouldn't have been such an epiphany to find that *Hirudo medicinalis,* the medicinal leech, could have more inherent dignity and complexity than a member of the county board of freeholders. Nonetheless, it was a revelation: All around us, even in our cities and suburbs, are creatures too strange, or often just too small, for us to recognize, and when I was able to look at life on their scale, through microscopes and macrophotography, and through the lens of scientific research, I found their lives fascinating. An alarming number of the creatures that caught my attention turned out to be what I have come to think of euphemistically as "spineless wonders." I mean invertebrates. Hence the title and the subject of this book.

An invertebrate is, of course, a creature without a spine, but in our scrupulously nonjudgmental modern world, why should this matter? Aristotle was the first to divide the animal kingdom into two major groupings. He called birds, fish, reptiles, amphibians, and mammals "Enaima," meaning "red-blooded animals." The insects, mollusks, jellyfish, corals, and other life-forms were "Anaima," animals lacking red blood. Though he was mistaken in using red-bloodedness as his criterion (earthworms, for one, are red-blooded), Aristotle had hit on a fundamental division: Animals in the first group are united by extensive similarities in the arrangement of organs, bones, and other features. Animals in the second group exhibit startling and often bizarre variety in their design and construction.

The French biologist Jean-Baptiste Lamarck modernized and expanded on Aristotle's idea, roughly two thousand years later. He pointed out that the one thing animals in the first group all possessed, and those in the second all lacked, was a backbone. So he coined the terms "vertebrate" and "invertebrate" and published them in his *Philosophie Zoologique* in 1809. Lamarck also published the first systematic examination of the invertebrate world.

But back to my original question: Why should this matter? Why should the normal reader care to join me on my invertebrate odyssey, into the realm of such indelicate creatures as the housefly, the tarantula, and the leech? Let me begin by appealing to baser instincts. As readers about nature, and as adventurers in the natural world, most of us secretly crave what Harvard sociobiologist E. O. Wilson has described as "a sweet sensation of horror, the shivery fascination with monsters and creeping forms that so delights today even in the sterile hearts of the cities." Our splendid language actually has a word for it: *Formication* is the feeling that ants or other creeping things are crawling over one's flesh. Much as we crave the thrill of horror movies ("bug" and "bogeyman" come from the same Welsh word, meaning "ghosts"), so too does good natural history writing give us the chance, vicariously and often covertly, to find pleasure in this sensation. My odyssey, by contrast, has been an unabashed wallow in the joy of formication.

Second, and still on the baser side, the people who make invertebrates their life work are often, to be frank, odd. To study lice, for instance, one British researcher kept them alive for two years in little pillboxes strapped under his socks, where they could feed on his blood. Most natural history writers overlook this aspect of the story, preferring to focus solely on the animals, perhaps a little embarrassed that their researchers might, after all, seem like geeks. But this is patronizing. The invertebrate zoologists I've met tend to be widely read and culturally aware. They know they often look eccentric to us, and, in truth, to each other. They even enjoy it. Gary Larson could be their house cartoonist. So this book sets out deliberately to celebrate the strangeness and wonder of both the invertebrate world and its attendant humans.

Apart from the entertainment value of these worlds, readers also have high-minded reasons to care about invertebrates.

The truth is that *we* are the geeks. The average person divides up the entire invertebrate world into two or three major categories: bugs, worms, jellyfish. Each of these words almost automatically causes us to look no further, and thus we write off the dominant life-forms of this planet. Natural history writers, television producers, and—until recently—zoos have tended to go back over and over to the familiar elephants and monkeys. But there are only 4,500 or so mammal species on the planet. There are, however, between ten million and thirty million invertebrate species. They represent more than 99.5 percent of all animal species. A spaceship visiting the blue planet would take *them,* not us, as the typical earthlings. They dominate even by sheer body mass. In one study, researchers calculated (don't ask how) that there are 178 pounds dry weight of animal tissue in an acre of Amazonian rain forest, and 93 percent of it consists of invertebrates—billions of invertebrates, mostly ants and termites, versus a few dozen birds and mammals.

By sheer number of species, and by the speed with which they can reproduce and evolve in the face of changing circumstances, invertebrates also wind up being more interesting than the common run of zoo animals. With invertebrates, nature always seems to be one-upping herself. She is not content merely to give us the quintessentially indolent sloth. She must also give us an even more idle moth species that breaks off its wings and resides patiently in the fur waiting to dine and lay its eggs on the sloth's weekly defecation. There are of course also mites on the moths on the sloths. Moreover, certain impatient beetles cohabit in the sloth's fur and are unwilling to live with their host's dilatory schedule. They sometimes crawl up the sloth's kiester to lay their eggs in the colon, giving the next generation a head start.

"Human beings are apt to regard their own personal structure as 'normal' and everything that differs from it as distinctly humorous," the English entomologist Miriam Rothschild has written. "It is difficult for them to realise that fleas breathe through holes in their sides, have a nerve cord below their stomachs and a heart in their backs; or that certain other arthropods lay eggs through their elbows, urinate through their heads and regularly practise virgin birth."

If the world of the invertebrates is often bizarre, it is also indispensable to life on Earth. We need these unseen invertebrate armies. Flies and moths are irreplaceable pollinators. Caterpillars are songbird fodder. Worms, not farmers, are the great plowmen of the earth, and if they ceased to till and fertilize the soil, or if insects no longer inhabited our fields and forests, we would soon starve. The bodies of invertebrates create the coral reefs and give rise to much of the life of the oceans. We are already beginning to see what might happen without them.

My ambition is that readers will finish this book somewhat dismayed but more often amazed by the marvels of the invertebrate world. I mean to evoke feelings somewhat like awe. If other natural history writing requires readers to be furtive about enjoying the sweet sensation of horror, my writing is more often furtive about what is lovely in the natural world. I have no knack for the sacred. My odyssey is about creeping sideways toward wonder. But the wonder is genuine.

I've found, in the course of this odyssey, that I am edging toward a healthier, more balanced perspective on the planet we share, however ambivalently, with the invertebrates. I haven't entirely given up pitching nature out with a fork. But I've learned to admire her as she comes crawling and slithering back. Now, when my other journalistic work requires me to do something truly odious, like interview a president or a prince, I console myself with thoughts of invertebrates, which I have come to recognize as the little things that really run the world.

Why Did God Make Flies?

Though I have been killing them for years now, I have never tested the notion, recorded in one collection of country sayings, that with a little cream and sugar, a fly "tastes very much like a black raspberry." So it's possible I'm speaking too hastily when I say there is nothing to like about flies. Unlike the poet who welcomed a "busy, curious fly" to his drinking cup, I don't cherish them for reminding me that life is short. Nor do I much admire them for their function in clearing away carrion and waste. It is, after all, possible to believe in the grand scheme of recycling without necessarily liking undertakers.

Among poets, I tend to side with Ogden Nash, who once wrote: "God in His wisdom / Made the fly / And then forgot / To tell us why."

A fly is standing on the rim of my beer glass as I write these words. Its vast, mosaic eyes look simultaneously lifeless and mocking. It grooms itself methodically, its forelegs twining together like the arms of a Sybarite luxuriating in bath oil. Its hind

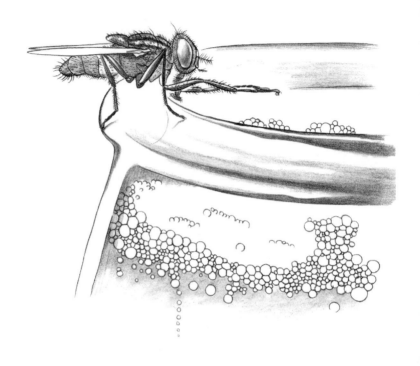

legs twitch across the upper surface of its wings. It pauses, well-fed and at rest, to contemplate the sweetness of life.

We are lucky enough to live in an era when scientists quantify such things, and so as I type and wait my turn to drink, I know that the fly is neither busy nor curious; the female spends 40.6 percent of her time doing nothing but contemplating the sweetness of life. I know that she not only eats unspeakable things, but that she spends an additional 29.7 percent of her time spitting them back up again and blowing bubbles with her vomit. The male is slightly less assiduous at this deplorable pastime, but one diligent researcher has reported that a well-fed fly may also defecate every four and a half minutes. Flies seldom trouble us as a health threat anymore, at least in the developed world, but they are capable of killing. And when we are dead (or sooner, in some cases), they dine on our corrupted flesh.

It is of course mainly this relentless intimacy with mankind that makes flies and particularly houseflies so contemptible. Leeches or dung beetles may appall us, but by and large they satisfy their depraved appetites out of our sight. Flies, on the other hand, routinely flit from diaper pail to dinner table, from carrion to picnic basket. They are constantly among us, tramping across our food with God knows what trapped in the sticky hairs of their half-dozen legs.

Twice in this century, Americans have waged war against flies, once in a futile nationwide "swat the fly" campaign, and again, disastrously, with DDT foggings after World War II. The intensity of these efforts, bordering at times on the fanatic, may bewilder modern readers. "Flies or Babies? Choose!" cried a headline in the *Ladies' Home Journal,* in 1920. But our bewilderment is not entirely due to greater tolerance or environmental enlightenment. If we have the leisure to examine the fly more rationally now, it is mainly because we don't suffer its onslaughts as our predecessors did. Urban living has separated us from livestock, and indoor plumbing has helped us control our own wastes, thus controlling flies. But if that changed tomorrow, we would come face-to-face with the enlightened, modern truth: With the possible exception of *Homo sapiens,* it is hard to imagine an animal as disgusting or improbable as the housefly. No bestiary concocted out of the nightmares of the medieval mind could have come up with such a fantastic animal. If we want to study nature in its most exotic permutations, the best place to begin is here, at home, on the rim of my beer glass.

In North America, more than a dozen fly species visit or live in the house. It is possible to distinguish among some of them only by such microscopic criteria as the pattern of veins in the wings, and so all of them end up being cursed as houseflies. Among the more prominent are the blue and the green bottleflies, with their iridescent abdomens, and the biting stable flies, which have served this country as patriots, or at least as provocateurs. On July 4, 1776, their biting encouraged decisiveness among delegates considering the Declaration of Independence: "Treason," Thomas Jefferson wrote, "was preferable to discomfort."

The true housefly, *Musca domestica*, of course does not bite. (You may think this is something to like about flies, until you find out what they do instead.) *M. domestica*, a drab fellow of salt-and-pepper complexion, is the world's most widely distributed insect species and probably also the most familiar, a status achieved through its pronounced fondness for breeding in pig, horse, or human excrement. In choosing at some point in the immemorial past to concentrate on the wastes around human habitations, *M. domestica* made a brilliant career move. The earliest known human representation of what appears to be a housefly is on a Mesopotamian cylinder seal from 3000 B.C.* But houseflies were probably with us even before we had houses, and they spread with human culture.

Like us, the housefly is prolific, opportunistic, and inclined toward exploration. It can adapt to either vegetable or meat diets, preferably somewhat ripe. It will lay its eggs not just in excrement, but in a rotting mass of lime peels, in bird nests, in carrion, or even in flesh wounds that have become infected and malodorous. Other flies aren't so flexible. For instance, *M. autumnalis*, a close relative, prefers cattle dung, and winds up sleeping in the pasture more than in houses or yards.

But while the housefly's adaptability and evolutionary generalization may be admirable, it raises one of the first great questions about flies: Why this dismaying appetite for abomination?

Houseflies not only defecate constantly, but do so in liquid

---

*Somewhat later, the ancient Romans used a poultice of mashed houseflies as a treatment for baldness. Flies are hairy, and the theory was that it might rub off on bald men.

form, which means they are in constant danger of dehydration. The male can slake his thirst and also get most of the energy he needs from nectar. But fresh manure is a good source of water, and it contains the dissolved protein the female needs to make eggs. She also lays her eggs in excrement or amid decay so that when they hatch, the maggots will have a smorgasbord of nutritious microorganisms on which to graze.

Houseflies bashing around the kitchen or the garbage shed thus have their sensors attuned to things that smell sweet, like flowers or bananas, and to foul-smelling stuff like ammonium carbonate, hydrogen sulfide, and trimethylamines, the products of fermentation and putrefaction. (Ecstasy for the fly is the stinkhorn fungus, a source of sugar that smells like rotting meat.) The fly's jerky, erratic flight amounts to a way of covering large territories in search of these scents, not just for food, but for romance and breeding sites. Like dung beetles and other flying insects, the fly will zigzag upwind when it gets a whiff of something good (or, as often happens, something bad) and follow the scent plume to its source.

Hence the second diabolical question about the housefly: How does it manage to fly so well? Why is it so adept at evading us when we swat it? How come it always seems to land on its feet, usually upside down on the ceiling, having induced us to plant a fist on the spot where it used to be, in the middle of the strawberry trifle, which is now spattered across tablecloth, walls, loved ones, and honored guests?

When we launch an ambush as the oblivious fly preens and pukes, its pressure sensors alert it to the speed and direction of the descending hand. Its wraparound eyes are also acutely sensitive to peripheral movement, and they register changes in light about ten times faster than we do. (A movie fools the gullible human eye into seeing continuous motion by showing it a sequence of twenty-four still pictures a second. To fool a fly would take more than two hundred frames a second.) The alarm flashes directly from the brain to the middle set of legs via the largest, and therefore the fastest, nerve fiber in the body. This causes so-called starter muscles to contract, simultaneously revving up the wing muscles and pressing down the middle legs, which catapult the fly into the air.

The fly's wings beat 165 to 200 times a second, and while this isn't all that fast for an insect, it's more than double the wingbeat of the fastest hummingbird, and about 20 times faster than any repetitious movement the human nervous system can manage. The trick brought off by houseflies and many other insects is to remove the wingbeat from direct nervous system control, once it's switched on. Instead, two systems of muscles, for upstroke and downstroke, are attached to the hull of the fly's midsection, and trigger each other to work in alternation. When one set contracts, it deforms the hull, stretching the other set of muscles and making them contract automatically a fraction of a second later. To keep this seesaw rhythm going, openings in the midsection stoke the muscles with oxygen directly from the outside (the fly has no lungs). Its blood (which lacks hemoglobin and is therefore colorless) meanwhile pumps fuel for the cells to burn at a rate 14 times faster than when the fly is at rest. Flies can turn a sugar meal into usable energy so fast that an exhausted fly will resume flight almost instantly after eating. In humans . . . but you don't want to know how ploddingly inadequate humans are by comparison.

Once airborne, the fly's antennae, between its eyes, help regulate flight, vibrating in response to airflow. The fly also uses a set of stubby wings in back, called halteres, as a gyroscopic device. Flies are skillful at veering and dodging; it sometimes seems that they are doing barrel rolls and Immelmann turns to amuse themselves while we flail and curse. But one thing they cannot do is fly upside down to land on a ceiling. This phenomenon puzzled generations of upward-glaring, strawberry-trifle-drenched human beings, until high-speed photography supplied the explanation. The fly approaches the ceiling rightside up, at a steep angle. Just before impact, it reaches up with its front limbs, in the manner of Superman exiting a telephone booth for take-off. As these forelegs get a grip with claws and with the sticky, glandular hairs of the footpads, the fly swings its other legs up into position. Then it shuts down its flight motor, out of swatting range and at ease.

While landing on the ceiling must be great fun, humans tend to be more interested in what flies do when they land on food, and so I trapped the fly on the rim of my beer glass. (Actually, I waited till it found a less coveted perch, then slowly lowered a mayonnaise jar over it.) I'd been reading a book called *To Know a Fly* by Vincent Dethier, in which he describes a simple way of seeing how the fly's proboscis works. First, I refrigerated the fly to slow it down and anesthetize it. Then I attempted to attach a thin stick to its wing surface with the help of hot wax. It got away. I brought it back and tried again. My son Jamie, who was then four years old, winced and turned aside when I applied the wax. "I'm glad I'm not a fly," he said, "or you might do that to me." I regarded him balefully but refrained from mentioning the ant colony he had annihilated on our front walk.

Having finally secured the fly, I lowered its feet into a saucer of water. Flies have taste buds in their feet, and when they walk into something good (bad), the proboscis, which is normally folded up neatly inside the head, automatically flicks down. No response. I added sugar to the water, an irresistible combination. Nothing. More sugar. Still nothing. My son wandered off, bored. I apologized to the fly, killed it, and decided to look up the man who had put me in the awkward position of sympathizing with a fly, incidentally classing me in my son's eyes as a potential war criminal.

Dethier, a biologist at the University of Massachusetts, turned out to be a gentle, deferential fellow in his mid-seventies, with weathered, finely wrinkled skin and a pair of gold-rimmed oval eyeglasses on a beak nose. He suggested mildly that my fly might not have responded because it was outraged at the treatment it received. It may also have eaten recently, or it may have been groggy from hibernation. (Some flies sit out the winter in diapause, in which hormones induce inactivity in response to shortened day length. But cold, not day length, is what slows down hibernating species like the housefly, and the sudden return of warmth can start them up again. This is why a fly may miraculously take wing on a warm December afternoon in the space between my closed office window and the closed storm window outside, a phenomenon I had formerly regarded as new evidence for spontaneous generation.) Dethier has spent a lifetime studying the fly's sense of taste, "finding out where their tongues and noses are, as it were." He explained the workings of the proboscis for me.

Fly taste buds are vastly more sensitive than ours, another reason to dislike them. Dethier figured this out by taking saucers of water containing steadily decreasing concentrations of sugar. He found the smallest concentration a human tongue could taste. Then he found the smallest concentration that caused a hungry fly to flick out its proboscis. The fly, with fifteen hundred taste hairs arrayed on its feet and in and around its mouth, was ten million times more sensitive.

When the fly hits paydirt, the proboscis telescopes downward and the fleshy lobes at the tip puff out. These lips can press down tight to feed on a thin film of liquid, or they can cup themselves around a droplet. They are grooved crosswise with a series of parallel gutters, and when the fly starts pumping, the liquid gets drawn up through these gutters. The narrow zigzag openings of the gutters filter the food, so that even when it dines on excrement, the fly can "choose" some microorganisms and reject others. A drop of vomit may help dissolve the food, making it easier to lap up. Scientists have also suggested that the fly's prodigious vomiting may be a way of mixing enzymes with the food to aid digestion.

If necessary, the fly can peel its lips back out of the way and apply its mouth directly to the object of its desire. While it does not have true teeth, the mouth of the housefly is lined with a jagged, bladelike edge, which is useful for scraping. In his book *Flies and Disease,* Bernard Greenberg, a forensic entomologist at the University of Illinois in Chicago, writes that some blowflies (like the one on the rim of my beer glass, which turned out to be an olive green blowfly, *Phormia regina*) "can bring one hundred fifty teeth into action, a rather effective scarifier for the superficial inoculation of the skin, conjunctiva, or mucous membranes."

Hence the final great question about flies: What awful things are they inoculating us with when they flit across our food or land on our sleeping lips to drink our saliva? Over the years, authorities have suspected flies of spreading more than sixty diseases, from diarrhea to plague and leprosy. As recently as 1951, the leading expert on flies repeated without demurring the idea that the fly was "the most dangerous insect known," a remarkable assertion in a world that also includes mosquitoes. One entomologist tried to have the housefly renamed the "typhoid fly."

The hysteria against flies early in this century arose, with considerable help from scientists and the press, out of the combined ideas that germs cause disease and that flies carry germs. In the Spanish-American War, easily ten times as many soldiers died of disease, mostly typhoid fever, as died in battle. Flies were widely blamed, especially after a doctor observed particles of lime picked up in the latrines still clinging to the legs of flies crawling over army food. Flies were not "dipterous angels," but "winged sponges speeding hither and thither to carry out the foul behests of Contagion." North American schools started organizing "junior sanitary police" to point the finger at fly-breeding sites. Cities sponsored highly publicized "swat the fly" campaigns. In Toronto in 1912, a girl named Beatrice White killed 543,360 flies, altogether weighing 212.25 pounds, and won a $50 first prize. This is a mess of flies, 108.7 swatted for every penny in prize money, testimony to the slowness of summers then and to the remarkable agility of children—or perhaps to the overzealous imagination of contest sponsors. The figure does not include the 2.8 million dead flies submitted by losing entrants. (The "swat the fly" spirit still lives in China. In 1992, Beijing issued 200,000 flyswatters and launched a major sanitation campaign under the slogan, "Mobilize the Masses to Build a City of No Flies.")

But it took the pesticide DDT, developed in World War II and touted afterward as "the killer of killers," to raise the glorious prospect of "a flyless millennium." The fly had by then been enshrined in the common lore as a diabolical killer. In one of the "archy and mehitabel" poems by Don Marquis, a fly visits garbage cans and sewers to "gather up the germs of typhoid influenza and pneumonia on my feet and wings" and spread them to humanity, declaring "it is my mission to help rid the world of these wicked persons / i am a vessel of righteousness."

Public health officials were deadly serious about conquering this archfiend, and for them DDT was "a veritable godsend." They recommended that parents use wallpaper impregnated with DDT in nurseries and playrooms to protect children. Cities suffering polio epidemics frequently used airplanes to fog vast areas "in the belief that the fly factor in the spread of infantile paralysis might thus be largely eliminated." Use of DDT actually provided some damning evidence against flies, though not in connection with polio. Hidalgo County in Texas, on the

Mexican border, divided its towns into two groups, and sprayed one with DDT to eliminate flies. The number of children suffering and dying from acute diarrheal infection caused by *Shigella* bacteria declined in the sprayed areas but remained the same in the unsprayed areas. When DDT spraying was stopped in the first group and switched to the second, the dysentery rates began to reverse. Then the flies developed resistance to DDT, a small hitch in the godsend. In state parks and vacation spots, where DDT had provided relief from the fly nuisance, people began to notice that songbirds were also disappearing.

In the end, the damning evidence was that we were contaminating our water, ourselves, and our affiliated population of flies with our own filth (not to mention DDT). Given access to human waste through inadequate plumbing or sewage treatment, flies can indeed pick up an astonishing variety of pathogens. They can also reproduce at a god-awful rate; in one study, 4,042 flies hatched from a scant shovelful, one-sixth of a cubic foot, of buried night soil. But whether all those winged sponges can transmit the contaminants they pick up turns out to be a tricky question, the Hidalgo County study being one of the few clear-cut exceptions. Of polio, for instance, Bernard Greenberg writes, "there is ample evidence that human populations readily infect flies. . . . But we are woefully ignorant whether and to what extent flies return the favor."

Flies thus probably are not, as one writer declared in the throes of the hysteria, "monstrous" beings "armed with horrid mandibles . . . and dripping poison." A fly's bristling unlovely body is not, after all, a playground for microbes. Indeed, bacterial populations on the fly's exterior tend to decline quickly under the triple threat of compulsive cleaning, desiccation, and ultraviolet radiation. (Maggots actually produce a substance in their gut that kills off whole populations of bacteria, which is one reason doctors have sometimes used them to clean out infected wounds.)

The fly's "microbial cargo," to use Greenberg's phrase, tends to reflect human uncleanliness. In one study, flies from a city neighborhood with poor facilities carried up to 500 million bacteria, while flies from a prim little suburb not far away yielded a maximum count of only 100,000.

But wait. While I am perfectly happy to suggest that humans are viler than we like to think, and flies less so, I do not mean to rehabilitate the fly. Any animal that kisses offal one minute and dinner the next is at the very least a social abomination. What I am coming around to is St. Augustine's idea that God created flies to punish human arrogance, and not just the calamitous technological arrogance of DDT. Flies are, as one biologist has remarked, the resurrection and the reincarnation of our own dirt, and this is surely one reason we smite them down with such ferocity. They mock our notions of personal grooming with visions of lime particles, night soil, and dog leavings. They toy with our delusions of immortality, buzzing in the ear as a memento mori. (Dr. Greenberg assures me that fly maggots can strip a human corpse roughly halfway to the bone in several weeks, if the weather is fine. Then they hand the job over to other insects.) Flies are our fate, and one way or another they will have us.

It is a pretty crummy joke on God's part, of course, but there's no point in getting pouty about it and slipping into unhealthy thoughts about nature. What I intend to do, by way of evening the score, is hang a strip of flypaper and also cultivate the local frogs and snakes, which have a voracious appetite for flies (flycatchers don't, by the way; they prefer wasps and bees). Perhaps I will get the cat interested, as a sporting proposition. Meanwhile I plan to get a fresh beer and sit back with my feet up and a tightly rolled newspaper nearby. Such are the consolations of the ecological frame of mind.

Little Suckers

One August afternoon some years ago in Medford, Massachusetts, a dog attacked a five-year-old boy and bit off his right ear. Until then, a child's ear had never been surgically reattached. The technical difficulties of this procedure were daunting: It was easy enough reattaching an artery to restore blood flow into the ear. But reconnecting the veins, said Joseph Upton, who performed the operation, was "like sewing together strands of wet toilet paper"—strands just a hundredth to two-hundredths of an inch in diameter.

During ten hours of work at a surgical microscope, Upton, a hand and microvascular surgeon at Children's Hospital in Boston, reconnected one artery and four veins. But within a few days the boy's ear began to discolor, turning blue and then purple as blood built up in it. What Upton needed was some means of gently relieving congestion while the veins healed themselves. The anticoagulant heparin wasn't doing the job. But it occurred to Upton that leeches might.

The leech, long ridiculed as a medicinal tool, was at that time attracting renewed interest among European microsurgeons. In fact, Upton had tried leeches himself some years earlier, as an army surgeon treating congested tissue grafts. Now, when he needed them again, he found them unavailable in the United States. A series of calls led him to a company called Biopharm in Swansea, Wales, where Lorna Sawyer agreed to drive a package of medicinal leeches (*Hirudo medicinalis*) the three hours to London's Heathrow Airport. Upton met the flight in Boston and, by the sheer good luck of finding a former patient in the federal bureaucracy, managed to wangle the delivery past a host of doubting customs agents.

The first leech soon fastened itself onto the upper crest of the boy's ear and arched its neck, which began to pulse subtly—a sign that bloodsucking had begun. Eight leeches and some days afterward, the boy went home with his ear intact, and newspapers and television newscasters hailed the triumph of the leech.

The source of Upton's leeches—and of most leeches used in medicine or research these days—was a sort of leech ranch, which was then operating in a disused Welsh steel plant. There, in darkened, temperature-controlled rooms, silent except for the whir and bubble of the aerator pumps, Biopharmers tended more than fifty thousand head of leech, of eight different species. Singly and in knotty, glistening clumps, the leeches clung to the sides of buckets, fish tanks, long, shallow vats, and a converted fiberglass dinghy. On top of each container, netting held down by strips of Velcro discouraged an alarming tendency of the animals to mosey off from these habitats. Biopharm, the brainchild of an eccentric but highly regarded American scientist named Roy Sawyer, was just getting its start then as a leech breeder and supplier.

Partly because of Sawyer, leeches today enjoy a scientific currency unrivaled since the leechmania of the early nineteenth century, when they were often applied fifty at a time to cure patients of every disorder from nosebleeds to gross obesity. But if the present interest in leeches is more restrained, it's also more promising in its potential for yielding new knowledge and new therapies.

Leeches nowadays do not merely rescue reattached ears, lips, fingertips, and other surgical grafts. They've also become a leading model for researchers seeking to understand how the nervous system works—how, for instance, some nerve pathways regenerate after traumatic injury. Leeches have also lately come under scrutiny as what Sawyer terms "a living pharmacopoeia." Several biochemicals evolved by the leech for the nasty business of getting and digesting blood have recently come on the market, as research tools. Some of them may someday help treat circulatory and other disorders. "Before I conk it," says Sawyer, his accent veering from South Carolina, where he was raised, to south Wales, "I want to develop one of them into a pharmaceutical drug." He believes that secretions from bloodsucking animals will eventually do for cardiovascular medicine what penicillin did for infections.

Sawyer is a mild man in his late fifties, soft-spoken, slightly distracted. He can turn any conversation to leeches, and his green eyes pop just perceptibly when he hits paydirt. According to his wife, Lorna, he was "born with a leech in his mouth." But he dates his obsession back only as far as his boyhood in the coastal swamps near Charleston, where every frog, turtle, and budding naturalist looked like dinner to a leech. Sawyer picked off the leeches and studied them, and one of his high-school reports included several species previously unknown to science, among them a turtle leech (*Placobdella translucens*) that was three inches long.

Sawyer naturally returned to the leech when he needed a subject for his doctoral thesis at the University of Michigan, and scholarly interest soon burgeoned into monomania. One night during their courtship, says Lorna, he took her out in party clothes and rubber boots to search a Florida swamp for a species that hadn't been seen since the 1870s. They failed to find it, but in the morning she heard an ecstatic whoop: a specimen of the leech had turned up inside one of her boots. She married him anyway. On another evening, sometime after, she heard an agonized cry from their living room, where Sawyer was in the habit of admiring a pair of ornate Chinese leeches. They had escaped, and the Sawyers eventually had to tear up the wall-to-wall carpeting to find them. His devotion to his subject (and her tolerance for it) is such that he was able to spend twelve years, five of them without gainful employment, compiling his definitive three-volume *Leech Biology and Behaviour,* published by Oxford University Press.

In fact, even back when leeches were held in contempt by the medical profession, Sawyer had a solid rationale for choosing them as his subject. Biology as taught in the United States had left him frustrated. "I felt that we were learning a mosaic," he says. "For sex determination, we'd study *Drosophila,* for physiology we'd study frogs, for genetics, bacteria. I thought there was more to be learned from studying one organism in detail than from parts of many." His American professors disdained this approach as a throwback to nineteenth-century biology. The British were more sympathetic, and at Swansea, Sawyer found a leech expert, E. W. Knight-Jones, to supervise his studies.

The biology and behavior of the leech were complicated enough to be interesting, but simple enough to allow detailed study. Moreover, because there are only about 650 species in the world, one person could master the field without spending his entire life doing taxonomy. Sawyer wanted to be able to move on to broader questions of evolution and relationships among species.

"I think the fact that they sucked blood also attracted me to them," he says. "I was interested on two levels. At first, leeches were just an obnoxious and repulsive creature that was absolutely fascinating. Later I became more focused on the biochemistry."

B ut it may be best to start, as Sawyer did as a boy, with mere natural history. In this sort of account, it's customary right about here to state that the animal in question has been sadly misunderstood by narrow-minded humanity. In truth, what most of us know about leeches is dead right: They're slimy, ill-mannered, hermaphroditic little brutes, and, frankly, some of them aren't so little.

The only serious misconception about leeches, which are annelids and closely related to earthworms, is the wishful idea that they are confined chiefly to steamy equatorial swamps. Most of us figure we'll never run into a leech unless we happen to take a dip in a tropical river, as Humphrey Bogart did in *The African Queen*. In truth, leeches thrive in alpine mountains and polar oceans. North America and Europe are the leech's favorite haunts, particularly in the north, where the glaciers of the last ice age carved out an abundance of watery habitat. "The Great Lakes are just perfect," Sawyer exults. One species, *Helobdella stagnalis*, lives on every continent except Australia. According to Sawyer, it will turn up even in your swimming pool, ferried in by birds, if you let the water stagnate.

A leech has both male and female sex organs. It doesn't fertilize itself, but copulates belly-to-belly with another leech. They

entwine in water or, in the case of medicinal leeches, on land. If they assume a head-to-tail position, it's possible, but not common, for both partners to be fertilized in a single encounter. The parent generally lays its eggs in a foamy cocoon, from which the hatchlings emerge. Some leeches fertilize their partners not by copulating, but by depositing spermatophores almost anywhere on the partner's body. These bore through the skin into the body cavity and thence to the ovaries. One species that reproduces in this fashion, *Haementeria ghilianii*, also carries its brood dangling from its belly, where they writhe like Medusa's hair.

The good news is that most leeches do not feed on people, or even mammals; many of them don't even suck blood. A typical freshwater pond in North America or central Europe might contain ten different leech species, which coexist by partitioning the food supply. The most common leeches feed on invertebrates. Some, equipped with jaws, gobble up whole midges, worms, even other leeches. Some insert a strawlike proboscis and suck the soft tissue out of snails and mosquito larvae. One American species (*Glossiphonia complanata*) dines in remarkable fashion on the fingernail clam. Like all leeches, it has suckers at both ends, for holding on and for walking from place to place in inchworm fashion. With its rear sucker, it attaches itself to the closed clamshell and waits there until the clam opens and puts out its foot. Then it dives inside headfirst. The clam snaps shut, but the intrepid leech manages to feed anyway, with its rear end hanging out. As the leech sucks up the clam's soft tissue, the clam loses muscle control and its shell opens. The leech, its belt loosened, finishes off the succulent remains.

Among bloodsucking leeches, some feed on flounder, some on octopus. (About a fifth of all leeches live in salt water.) Some live in caves and feed on bats. Some can penetrate the skin of a hippopotamus or an elephant, while others can feed only through soft mucous membranes. Herodotus described the spectacle of Nile crocodiles basking while birds rummaged about in their open mouths, picking leeches off their gums. Ducks may also get leeches in their throats and nasal passages, and can be seen sneezing or scratching at their bills to dislodge them. Even humans are vulnerable to this sort of invasion, and both Pliny's *Natural History* and the *Talmud* warn of nasal leeches (*Limnatis nilotica*) in drinking water. In central Africa, a related species still sometimes causes death by anemia and suffocation.

The idea of using leeches medicinally was an outgrowth of the belief that disease was caused by an excess of corrupt blood or by an imbalance in bodily humors. The leech doesn't merely consume five to ten times its weight in blood during a half-hour feeding; the wound it creates bleeds for hours afterward, for reasons scientists have only recently begun to understand. The earliest written reference to the medical application of leeches, by a Greek in the second century B.C., advised their use in treating venomous bites. But medicinal leeching apparently originated much earlier in India.

As he pursued his studies, Sawyer speculated that there might be some kernel of truth to leeching, some connection between mammal and leech bordering on symbiosis, bred not in the bone but in the blood. It was a crackpot idea, particularly in light of leeching's recent history. Until the nineteenth century, leeching had generally been reserved, as one ancient Indian medical treatise put it, "for women and timid persons" and for delicate parts of the body—for instance, around the eyes and in the mouth. More radical forms of bloodletting—with lancets, fleams, scarificators, and cups—were generally preferred. But when leeching achieved its brief heyday, from about 1820 to 1850, the popular enthusiasm for it bordered on madness.

Leeching's chief proselytizer François Broussais, a French physician, once treated his own indigestion over an eighteen-day period with fifteen applications of fifty to sixty leeches each. Dresses decorated with embroidered leeches, called "robes à la Broussais," were one season's fashion, and a contemporary author estimated that medicinal leeches were purging 363,000 liters of French blood annually. In one case, a young French woman retreated to her room with fifty leeches to attempt suicide; like almost everyone else, she overestimated the powers of the leech.

Demand was so great that the medicinal leech almost became extinct in Europe. Leech-gatherers plied their trade simply by walking around in water and allowing leeches to attach themselves to their bare legs. A person could harvest twenty-five hundred leeches a day from a good pond, and in 1824, the yield from this primitive technique allowed Germany to ship five million leeches to England in a single consignment.* When the natural supply gave out, the French and Germans took up leech farming, feeding the young according to a reported formula of ten elderly horses per hectare of breeding area. Buyers made forays into eastern Europe, and their depredations were so thorough that the Russian government imposed tariffs and established a closed season on leech-gathering. (The medicinal leech is still threatened in Europe, partly because of habitat loss through swamp drainage. Commerce in medicinal leeches collected in the wild is currently banned by international agreement.)

But it was not scarcity so much as the therapeutic excesses of leechmania that eventually dragged the leech down into utter disdain. The death of Russian novelist Nikolai Gogol was merely one instance: Already debilitated from terminal illness and suffering from a nosebleed, Gogol was doused in cold water while six leeches fed at his nostrils. An eyewitness later wrote, "When the leeches were attached he repeated, 'Take the leeches off, take the leeches off my mouth,' and he tried to reach them with his hands." The patient had to be restrained during prolonged sessions of leeching. "All this probably helped him die much faster."

Despite such horror stories, leechmania had some beneficial effects. When modern science began tentatively to reconsider the

---

*In the poem "Resolution and Independence," written in 1802, Wordsworth meets an old man "on the lonely moor" who testifies that leeches had become scarce in England:

> He with a smile did then his words repeat;
> And said that, gathering leeches, far and wide
> He travelled; stirring thus about his feet
> The waters of the pools where they abide.
> "Once I could meet with them on every side;
> But they have dwindled long by slow decay;
> Yet still I persevere, and find them where I may."

leech, it was partly because nineteenth-century research suggested tantalizing possibilities. John Nicholls, a neurobiologist at Harvard University in the early 1960s, established the leech's potential as an experimental animal. He was interested in how nerve cells connected and communicated, and it had been known since the 1890s that the medicinal leech had large and unusually accessible nerve cells suitable for such a study. With the leech, Nicholls could pick out a particular neuron under the microscope and study it in detail. He could penetrate the cell membrane with microelectrodes and measure the cell's electrical properties.

The leech's neurons weren't merely large, but also relatively few in number—about ten thousand rather than the estimated ten billion in the human nervous system. And where human nerve cells occur in compact masses, leech neurons are generally spread out in a single layer. It was thus possible not just to study a particular cell, but to map it and find it again for subsequent studies. Getting to know the entire animal was also easier because the leech is divided into thirty-two segments, with considerable repetition from one segment to the next. Its central nervous system consists of thirty-two neuronal bundles, or ganglia, tied together by connective nerves, and much of what researchers learn about the two hundred neuron pairs at ganglion 12 applies equally to the two hundred pairs at ganglion 22. Despite this simplicity, the leech does many of the complicated things that other animals do, including rhythmic activities like walking and swimming.

Gunther Stent, who worked with Nicholls, had the idea of using the leech to make a start-to-finish analysis of how the nervous system controls a single complex behavior. No one had ever accomplished this in any animal, despite its obvious utility. Nature is conservative, and biological mechanisms tend to be repeated from one species to another. Deciphering the mechanisms that control chewing, walking, or swimming in one simple species can thus provide clues about how much more complex species, including humans, accomplish these same activities. This information could lead to an understanding of why the human nervous system sometimes fails and, eventually, how to fix it.

Stent, while hardly indifferent to benefits that might result from his work, was more interested in the leech itself. He decided to concentrate on swimming, despite warnings that this would prove too complicated for analysis even in the leech. Leeches are wonderful swimmers, flattening out their bodies and fluttering through the water with the rhythmic up-and-down undulation of sea monsters. Leonardo da Vinci sketched them for his studies of locomotion. They can swim upside down. They can skim along the contours of a rock. When hungry, they are quick to swim toward the source of a disturbance. They can tread water next to your leg, while their front end probes delicately for a suitable place to fasten on.

In Stent's lab at the University of California at Berkeley, and in laboratories later established by his postdoctoral students Bill Kristan at the University of California at San Diego and Otto Friesen at the University of Virginia, researchers began to explore the inner workings of nervous system control. They started with the muscles that produce swimming and the motor neurons that control these muscles. In each of the leech's body segments, muscles running lengthwise in the belly and the back pull against one another, one side contracting while the other relaxes. This sequence is coordinated down the length of the body, and the closer a segment is to the tail, the later the contraction-relaxation sequence occurs, producing the characteristic sea-serpent pattern of troughs and crests.

To determine whether this activity was largely reflexive—the product of a simple circuit from sensory cell stimulus in the body wall to motor neuron to muscle—Kristan removed the central nervous system. Stretched out in a dish, with the ganglia visible like beads on the central cord, it resembled a necklace. In 1975, Kristan was able to get this isolated cord to produce the rhythmic pattern of back-and-belly motor neuron activity, thereby demonstrating that the central nervous system, not mere reflex, generates the swimming rhythm.

Researchers began to decipher a complicated hierarchy of control mechanisms in the central nervous system. One step up from the motor neurons in the chain of command was a group of so-called interneurons—nerve cells connected only to other nerve cells—which seemed to generate the pattern for motor neuron activity, beating out the rhythm like the drummer in a band. And above the oscillating interneurons was a swim-initiating neuron present in about half the ganglia.

Friesen's group was studying small structures in the leech's skin, called sensilla, attuned to water waves, a stimulus that causes hungry leeches to swim. They hoped to find a direct connection between the sensory neurons in these structures and the swim initiator. Instead, they found one final layer in the hierarchy, a group of trigger cells in the fused ganglia of the first four segments, which is what passes in the leech for a brain.*

By 1986, leech neurobiologists had identified every level of the circuit—sensory neurons to trigger cell to swim-initiating

---

*On the subject of the brain, Roy Sawyer's *Leech Biology and Behaviour* quotes admiringly from Nietzsche:

> "In the true knowing-knowledge there is nothing great and nothing small."
>
> "Then thou are perhaps an expert on the leech?" asked Zarathustra; "and thou investigatest the leech to its ultimate basis, thou conscientious one?"
>
> "O Zarathustra," answered the trodden one, "that would be something immense; how could I presume to do so!
>
> "That, however, of which I am master and knower, is the *brain* of the leech; that is *my* world!"

interneurons to oscillating interneurons to motor neurons to muscles—and had established that other nerve cells modulate swimming by means of serotonin, a neuronal secretion. It was a triumph of basic research. As a sign in the office of one leech neurobiologist puts it, "*felix qui potuit rerum cognoscere causas*": "Happiness is understanding how things work."

Stent also focused his attention on leech embryology, asking how and why a nerve cell acquires its anatomical, biochemical, and electrical properties. Stent pointed out that nineteenth-century researchers had already established the value of the leech as a tool of embryology: It was the leech that first effectively challenged Ernst Haeckel's famous "biogenetic law," which cryptically decreed that "ontogeny recapitulates phylogeny." Haeckel, a leading exponent of Darwinism, believed that in its earliest development an animal passed through the stages of its evolutionary history, beginning with the undifferentiated tissues of some remote, spongelike ancestor. By studying how cell lineages developed in the leech embryo, Charles O. Whitman of the Woods Hole Marine Biological Laboratory in Massachusetts demonstrated in the 1880s that recapitulation had nothing to do with it. Instead of being amorphous or undefined, each identified cell and its descendant cells played a specific, predestined role in later development.

To get his studies off the ground, Stent turned in 1976 to Roy Sawyer, who was then pursuing his leech studies full-time at his home in Wales. Sawyer had evaded academic employment ("Traditional routes have always sort of defied me," he says), but he agreed to join Stent's group at the University of California at Berkeley, where he set about identifying suitable species for embryological studies. He came up with two candidates, a dwarf leech native to California that feeds on snails (*Helobdella triserialis*), and the giant Amazon leech (*Haementeria ghilliani*), which feeds on the blood of mammals. It also grows to eighteen inches, but somehow no sightings had been reported in this century.

Sawyer was soon on his way to French Guiana, where he began collecting leeches by the time-honored method of walking bare-legged in coastal marshes. The thirty-five giant Amazon leeches that Sawyer brought back and bred up to a working population of ten thousand were to provide Stent and other researchers with the material for years of valuable developmental studies. But Sawyer also had his own reasons for being interested in *H. ghilliani*. He had some questions he wanted to answer about the leechy business of getting blood.

The medicinal leech and the giant Amazon leech represent two divergent approaches to the surgical problem of penetrating skin. *H. medicinalis* is equipped with three cutting plates, each with sixty to one hundred teeth. Under a microscope, these jaws look like the blades of a circular saw. But in fact they cut into the flesh with a reciprocating motion, and because they're arranged radially, they open the wound in three directions at once. (Leech fanciers like to say that the scar from the bite resembles the hood ornament on a Mercedes.) By contrast, the giant Amazon leech penetrates the skin with a long, threadlike proboscis.

Bloodsucking leeches "need to be somewhat devious in order to feed," says Sawyer. They deploy "a cocktail of secretions," possibly including an anesthetic, to evade detection and subvert a victim's blood biochemistry to their own furtive ends. The medicinal leech secretes a substance called hirudin, which was the world's first known anticoagulant when researchers discovered it in 1884. But hirudin seems to work mainly in the leech's own gut, keeping the meal liquid during digestion so the animal doesn't suddenly become stiff as a Cheez Doodle.

To get the blood out of the victim in the first place, the leech relies more heavily on a chemical that dilates the veins. It also uses a so-called spreading factor, which breaks down the "cement" that binds cells together and disperses the leech's other salivary secretions well beyond the bite itself. These and other chemicals are what kept the Massachusetts boy's ear draining, even between leeches, after microsurgical reattachment. In his book, Sawyer speculates that the spreading factor may partly account for the persistent belief that leeching can cure disease: Many bacteria are protected by a capsulelike coating, but the spreading factor dissolves this protection, making the bacteria vulnerable to the host's immune defense mechanisms. In experiments on mice and guinea pigs, the spreading factor has killed streptococcal bacteria, which can cause virulent infections. But the leech's bite may also be a source of infection. Researchers in Bristol, England, have reported six cases of wound infection in patients to whom leeches had been applied to relieve congestion in the veins. The source was a bacterium that lives in the gut of the medicinal leech and apparently gets passed to a leech's victim along with the salivary secretions.

This bacterium, *Aeromonas hydrophila,* is one of the most remarkable aspects of the leech's culinary biochemistry: A leech sometimes takes six months to digest a meal, and yet the blood in its crop remains not just liquid but also free of putrefaction. The leech produces hardly any digestive enzymes. Instead, it relies on *A. hydrophila* to digest the blood meal, and this bacterium puts out an antibiotic to kill other bacteria that might cause putrefaction.

It occurred to Sawyer that because proboscis-feeders and medicinal leeches evolved separately, they might have different biochemical cocktails. He focused on the giant Amazon leech because it's one of the few proboscis-feeders that feed on mammals. And in fact, Sawyer and Andrei Budzynski, a researcher at Temple University, found that the giant Amazon leech had an anticoagulant unlike any other then known, which they named hementin. Whereas hirudin neutralized an enzyme, called thrombin, needed to form the protein fibers of a clot, hementin attacked fibrinogen, a blood plasma protein, not only preventing coagulation, but also dissolving clots after they formed. Unlike other anticoagulants, hementin was also indifferent to substances in human blood plasma that normally prevent proteins from being dissolved. The potential for using hementin to treat blood clots was obvious.

After Sawyer helped Stent get his program under way, he and Lorna went back to Wales, where her family lives. He finished writing his book. Then one morning he woke up and realized that nobody was going to develop hementin's potential unless he did it himself. He created Biopharm, and Lorna and her parents came to work. Twelve years later, it moved out of the steel plant, to a Victorian manor house with twenty-five acres of land near Swansea. Sawyer has built what he terms "a much more upmarket leech breeding facility" on the grounds. In place of his windowless, undecorated office at the steel plant, Sawyer now has a view of the Loughor River. There is still no rug on the floor (it would only be another place for runaway leeches to hide). But Biopharm also has offices in four other countries, as well as informal ties to scientists from Moscow to La Jolla. Every year, it ships about fifty thousand leeches to researchers and microsurgeons in two dozen countries.

Much of Biopharm's revenues come from the sale of leech chemical extracts, among them hirudin, hementin, and the spreading factor, which Sawyer calls Orgelase (Welsh for "from the leech"). Eventually, Sawyer aims to mass-produce the chemicals with the help of genetic engineering. Hementin is currently in production development, with human testing to begin in two or three years. Pharmaceutical companies phone constantly with questions, and Sawyer finds himself in the felicitous position of earning a living while thinking about nothing but leeches twenty-four hours a day, seven days a week. In his spare time, he has opened a leech museum at the Biopharm office in Charleston, South Carolina.

One afternoon not long ago, Sawyer took time out to talk about the leech's brilliant prospects. He was a man in love with his subject, capable of comparing the slime on the leech's skin to birdsong—a means by which one animal recognizes another of its own species. As he talked, part of him maintained a cautious scientific reserve: He preferred not to discuss a report that a salivary extract from the giant Amazon leech inhibits cancer in mice, because the mechanism is unknown and application in humans is remote at best. But part of him could not help being expansive on the leech's behalf: He was optimistic, for instance, about research indicating that the spreading factor may someday provide a treatment for certain forms of glaucoma.

Then Sawyer noticed a blood blister on his visitor's palm. His eyes widened. "Uh, we can fix that," he offered. The visitor, having just been escorted past fifty thousand leeches, declined with unseemly haste. Better to wait—barring ear reattachment or fingertip replantation—for the day when Sawyer and the medical world can work the full spectrum of leech magic. Better to wait, thanks all the same, till they can do it without firsthand reference to the leech itself.

Empire of the Fire Ant

One sure sign that some passing phenomenon has insinuated itself immortally into modern American culture is that somebody tries to build a tourist attraction around it. It doesn't matter how awful the object in question may be. In Massachusetts a few years ago, for instance, people in Fall River wanted to set up a museum for Lizzie Borden, the ax murderer, "as long as we can do it tastefully." In almost the same spirit (taste being frankly less relevant), the town of Marshall, Texas, has put itself on the map with an annual celebration of the imported red fire ant, a nasty little invader with an almost mythic reputation as a deerslayer, a mankiller, a molester of costly agricultural machinery.

Among other gala events, the festival features a fire ant roundup. What you do, see, is kick a fire ant mound, and then, as the ants come seething up, you direct them into a bottle, which is kind of like smashing a hive and trying to round up the bees. There is also a fire ant mating-call contest (best left to the imagination), a Miss Fire Ant beauty competition, and a theme song: "A fire ant'll bite everything in sight / without so much as a by-your-leave. / They'll scare your kids and make your dogs uptight / and make you question all that you be-*leeve*." A civic-minded chemical company sponsors the event and has at times used jugs of its fire ant killer to spell out the words "FIRE ANT FESTIVAL" on a football field. The message, subtly incorporated into the theme song, is that without this product the fire ants'd "get us all." Meanwhile, why not have some fun?

As a cultural icon, the fire ant qualifies even more eminently on a second count, which is that politicians have long recognized it as an intensely emotional issue, and they've pushed it hard. Since fire ants arrived accidentally from South America sometime before World War II, they have spread, in the podium-thumping rhetoric of old campaign speeches, like Sherman's army, like a biblical plague, like the red tide of Communism, across 260 million acres in eleven states, from Florida to Texas. At the moment, they are threatening to break out of the South and provide demagogic raw material for ambitious politicians in New Mexico, Arizona, and California. (Imported fire ants have appeared for the first time along the Rio Grande, which is expected to serve as their highway past the dry terrain of west Texas. But they also hitchhike west by truck almost daily, and they have set up shop briefly in the Phoenix and Santa Barbara areas before being exterminated.)

History suggests the politicians should be wary. For decades, southern governors and congressmen rode the fire ant into office, and legislated pork-barrel programs to eradicate it once they got there. It is a tempting issue, because fire ants bedevil such an array of constituents: Their mounds, up to eighteen inches high, are a form of agricultural acne and a blemish on suburban lawns. Forty mounds to the acre is a typical infestation, but in the last decade a new improved version of the fire ant has achieved densities of four hundred mounds per acre across large areas of Texas. The ants can damage crops, and their mounds bend and foul costly field machinery. Though no study has ever authenticated it, farmers have charged for decades that fire ants sometimes kill newborn calves. In Texas, they swarm over fawns bedded down by their mothers, and when the fawn tries to lick them off, they sting it even inside its stomach, where autopsies have found them in the hundreds. Some scientists fear that, at high densities, fire ants may imperil whole populations of small animals.

Fire ants also of course sting people, three million to five million Americans a year by one estimate. They will climb any foot left for a few seconds in the wrong spot, and when the advance forces get to about the knee the whole gang stings en masse. "It's very embarrassing to be at a church social," one expert writes, "and have to remove your pants due to fire ants." Not content with having made the outdoors less habitable, fire ants can also swarm over electrical equipment and shut down traffic signals, airport runway lights, and those essentials of couch-potato civilization, air conditioning and television. They get into kitchens and clothes hampers. They sometimes sting incapacitated hospital patients in their beds.

Fewer than 1 percent of sting victims suffer an allergic reaction requiring medical attention. But toddlers who've fallen on fire ant mounds have sometimes died from stings, as have allergic adults. These cases are extremely rare and hard to document, a drawback that does not keep worried parents from hyperventilating over the imagined peril in their own backyards. Depending on whom you believe, fire ants are a genuine red peril or an overrated object of mass entomophobia.

Over the years, fire ant horror stories have taken on a life of their own in political speeches, newspaper headlines ("Uncontrolled fire ants ravage South," "Texas losing ground in war with killer ants"), and the jacket copy of at least one novel: "Violence erupted as the Fire Ants swarmed over the peaceful valley, killing every living thing."

These stories have often been reminiscent of a nightmarish, mid-twentieth-century short story, "Leiningen Versus the Ants," read by generations of high school students and helpfully categorized by their earnest English teachers as an instance of the literary subgenre "man against nature." In it, a Brazilian plantation owner supremely confident in the power of the human brain prepares to resist the advance of a twenty-square-mile army of "ants, nothing but ants! And every single one of them a fiend from hell; before you can spit three times they'll eat a full-grown buffalo to the bones."

With moats, an insecticidal spray, and the help of four hundred "peons," the plantation battles the sea of "thumb-long" ants for days, eyeball to "brilliant, cold" eyeball, shovels to "razor-sharp mandibles," before Leiningen ultimately triumphs, having been chewed to the bone in no more than one or two places.

The trouble with the story is that, in real life, Leiningen lost. In truth, there has never been a good way to eradicate *Solenopsis invicta*—to use the imported fire ant's dryly objective scientific name, which means "unvanquished."

Politicians, visions of peons and moats dancing in their heads, have never quite taken the point. Not along ago, for instance, a Texas commissioner of agriculture was campaigning for reelection. In search of a good television visual, he stuck his hand into a fire ant mound and held it there while expounding on his promise to "whip" the fire ant with chemicals. Then he did it again for another camera crew, a gesture so contrary to once-bitten-twice-shy logic that Texans rose up in righteous itching horror to elect a more commonsensical opponent.

The fire ant has proved itself without equal in its ability to make fools of politicians, bureaucrats, and ingenious inventors who thought they could whip it. It is chapter-and-verse environmental history. In the 1950s, the U.S. Department of Agriculture declared fire ants a menace to crops in the Southeast and sprayed the pesticide heptachlor across twenty-seven million acres, killing fire ants—as well as birds, reptiles, mammals, and fish, including farm animals and pets. The program became one of Rachel Carson's chief targets in *Silent Spring.*

The USDA responded with "the perfect pesticide," mirex, which it said would eradicate the imported fire ant without harming wildlife or leaving residues in farm products. After sixteen years in which bombers sprayed it over 130 million acres, researchers persuaded the government that mirex was a potent killer of crustaceans and a possible carcinogen, which bioaccumulates as it moves up the food chain, concentrating wonderfully in human fat tissue. Thus ended what ant expert and sociobiologist E. O. Wilson called "the Vietnam of entomology." The analogy seems apt even now, when you can walk into a taxidermist's shop in a place like El Campo, Texas, and hear a soybean farmer blaspheme the "stupid do-good bastards" who killed mirex: "We almost had 'em eradicated. One more plane going-over would've done it."

With the fire ant, alas, things never turn out to be that simple. It now appears that the pesticide program actually helped *spread* imported fire ants—at a cost to taxpayers of $172 million over twenty-four years. By indiscriminately killing all ant species, including some that prey on fire ant queens, pesticides like heptachlor and mirex seem merely to have cleared the field for the invader. The imported fire ant, a weed species with prodigious reproductive potential, quickly colonized all the newly vacant niches in the ecosystem. An additional irony (and ironies abound with fire ants) is that while fire ants are often an agricultural pest, they were never so demoniacally destructive as official lore suggested. At the USDA fire ant research center in Gainesville, Florida, officials are quick nowadays to supply a list of Florida's agricultural pests, on which the imported fire ant ranks twenty-second in economic impact, down with chinch bugs and leafhoppers. Because fire ants happily eat other insect pests, notably the sugarcane borer and the boll weevil, farmers in some areas actually regard them as beneficial.

The failed war on fire ants produced widespread frustration. But American ingenuity rose to the occasion with a host of environmentally safe new ways to kill fire ants. The McCoy Ant Stomper, for example, used a windmill to drive a small wheel around a circular track, killing any ant dumb enough to get in its path. Another device, the Antser, was a kind of sit-down lawnmower with an oversized eggbeater underneath for literally whipping the fire ant mound (an invention with enormous potential as a Fire Ant Festival equivalent of the bucking bronco contest, the test being how many seconds a person can remain seated as the fire ants rush up the eggbeaters to sting their tormentor). The most popular remedies were the ones that gratified what Mark Trostle of the Texas Department of Agriculture calls "the KCK factor"—the human need to "kick, cuss, and kill." Entrepreneurs sold devices to suck fire ants out of the mound and to incinerate them in their underground tunnels. Among home remedies, dry grits were said to swell up in the ant's digestive tract and "bloat them to death." (Scientists demur: Fire ants have hairs in their throats that filter out morsels at the submicron level, including individual grits.) Another home remedy, requiring a certain loose knowledge of fire ant territoriality, was to shovel one mound on top of another, so the rival colonies could sting each other to death. Less erudite but arguably more gratifying was the practice of drinking Lone Star beer and urinating on the mound.

But the failure of mirex had one genuinely beneficial result: It produced a belated interest in learning how fire ants really live, if only to kill them more effectively. In the last decade, the imported fire ant has become one of the most carefully studied of the fifteen thousand ant species in the world, researched, with the help of advanced scientific machinery, down to the sequencing of its genes and the gas chromatography of its pheromones.

Some of the most sophisticated work has come from a Florida State University researcher named Walter Tschinkel, whose preferred research instrument is a short-handled shovel. Tschinkel, lean, in his mid-fifties, with thick brown hair and deep-set, ironical eyes, takes a pure, tinkering delight in his ants. The world inside an ant mound is his Serengeti. "When you focus down to the world of an ant, to that scale," Tschinkel told me, "you see that they're really very complex animals. They're beautiful in their body form. They're lovely in their behavior. They are masterpieces of miniaturization. They would make any Silicon Valley person very proud." It vexes him a bit when the only question people ask is if he's come up with a surefire way to kill the object of his fascination ("Put the ant on a brick," he advises, if pushed, "and take a hammer . . .").*

Tschinkel's laboratory has specialized in the use of elegant low-budget techniques to answer complex questions about how fire ants live. For one study, he and an associate, Sanford Porter, needed to separate each of thirty-four colonies averaging 38,000 fire ants apiece according to individual ant size—the sort of job the gods would give Sisyphus in hell. Tschinkel's solution

---

*Shortly after Tschinkel told me this, I happened to stop at a lakeside camp near Tallahassee, where I met a man sitting outside his trailer watching fire ants climb onto a brick and methodically crushing them. He told me his name was Robert Burns, and I would have accused him of kidding me, except that I am reasonably sure he had never heard of the poet. I could almost hear the refrain "wee sleekit cowrin, tim'rous beastie . . . *whap!*"

was to use a stack of successively finer geological sieves. With the help of "negative phototaxis" (the urge to run away from a 60-watt light bulb shining directly overhead) and some judicious banging, they got the ants to crawl down through the levels until the mesh size became too small to go any further. Result: neat division of the colony into six sieves, each containing ants of uniform size. They used the technique to study how different-size ants affect the efficiency of the colony as it goes about its business.

Tschinkel has done much of his fieldwork at a new housing tract a short drive from his office in downtown Tallahassee, and we drove there one day to visit. Fire ants, like other weed species, prefer areas disturbed by development. One reason they've spread so rapidly through the South is their knack for colonizing nurseries and sod farms and traveling with the merchandise to every new construction site. In the middle of a weedy field, Tschinkel quickly found a mound and sliced into its sunny side with his shovel, an eighty-year-old hand-me-down. "Let's see if we can find a queen," he said, breaking up clumps of earth and spreading the fire ants out across a flat piece of ground. He paused occasionally to whip the ants off his hand before they could sting.

Queens mate once, early in their lives, in a gaudy nuptial flight up to four hundred feet in the air. The males promptly die (presumably happy). Inseminated females by the hundreds or thousands find their way to sites like this, typically within a half-mile of their birthplace, and get on remorselessly with the main business of living, which is nonstop reproduction. They shed their wings, dig in, and try to establish a colony. Most of them fail, starving to death or becoming tidbits in the diet of kingbirds, armadillos, or other ants.

If a female survives, she quickly lays her first clutch of perhaps sixty to ninety eggs. Alone in the burrow, without any outside food, she will use up half her body weight in the month before her first workers are ready to tend her. The odds that any queen will survive this perilous founding stage on her own are minute. (Tschinkel once calculated that, out of a thousand new queens that dug in at his study site, only one lived to head a successful colony.) Some queens choose to live in a communal nest with ten to fifteen rival queens who are complete strangers. Ultimately, the worker ants will execute all but one of them, possibly sending their own mother to the chopping block, to produce the standard single-queen colony. This is the sort of survival issue that intrigues Tschinkel: Does a female give up her independence because her chances of passing on her genes are better in a group, even considering the likelihood of execution?

To establish her future empire, the surviving queen's workers next set out to become kidnappers. Stealing the brood of other colonies is the fastest way for a queen to build up the largest possible army of workers. The colony that starts out with the most workers—presumably the one with the largest number of cooperating queens—has a tremendous advantage. "That's the one that's going to win the race for space," said Tschinkel. "It's what General Nathan Bedford Forrest said about winning a battle. 'You've got to get there the firstest with the mostest.' "

Tschinkel once witnessed a "mega-raid" at his study site, involving eighty colonies and lasting thirty-five days. It began in pandemonium, with everybody stealing everybody else's eggs and larvae across a hundred-foot-wide battlefield. "It's sort of the equivalent of carrying your own weight in a broiling sun at a full run from here to Quincy, the next town, which is twenty-five miles away," Tschinkel said. Gradually, weaker colonies collapsed into larger ones—presenting another intriguing sociobiological question: The early offspring are all sterile. So what causes the losing workers at some point to abandon their mother, grab the brood, and throw in their lot with the winning colony—in which they can have no genetic stake? Tschinkel found that the losing queen may actually tag along after her fickle offspring. Their presence can help her gain entry to the victorious colony. There, with the luck and good timing to survive the regicides, she may actually end up on top.

"There she is!" Tschinkel said, pointing to a tangle of ants from the mound he had just opened up. In the middle, the half-inch-long queen seemed to float on her jostling attendants like a bloated Egyptian idol in a Cecil B. DeMille spectacular. The workers, innocuous-looking creatures less than an eighth-inch long, played pattycake with their antennae on her flanks, picking up chemical messages and attending to her needs.

"The term 'queen' gives entirely the wrong impression," said Tschinkel, plucking her out of the tangle. She's not a ruler, he said, but a reproducer, a fire ant assembly line through which the workers constantly sluice food to make eggs. He pointed to her tiger-striped hind end, or gaster, where the swollen, cream-colored ovaries were visible through the stretched-out intersegmental membranes. "She can be up to seventy-five percent ovaries by weight," he said. "She's really a machine. It's just boring. All the time, eggs, eggs, eggs." From her single mating, he said, she parcels out the sperm to produce sixty to two hundred eggs per hour, a daily output equal to her own body weight—and she keeps it up for perhaps six years. "She will

have approximately three million daughters," plus a small annual crop of winged males to assist in founding new colonies.

It was a little difficult to grasp just how phenomenally boring this must be, so I made some anthropomorphic calculations: A single queen weighing twenty-four milligrams will produce up to eight hundred grams of offspring annually. This is roughly the equivalent of a 120-pound woman giving birth to 500,000 eight-pound babies a year. I looked at the queen and thought, "You can do it, honey, just keep panting."

In Tschinkel's hand, separated from her nestmates, the queen was listless and immobile. But as we watched, eggs continued to pump out, tiny spheres sliding down her stinger. Automatically, she anointed them with her venom, which conveys chemical messages and may also have an antibacterial effect. "Poor thing," he said, putting her back in the mound, to be caught up again in a sea of attendants.

Killing the queen is the ultimate object of most fire ant research, and it has never been an easy task. The colony's nest is almost entirely below the surface, a complicated hive of chambers arranged like subterranean apartments around multiple elevator shafts, often running six feet underground. The colony moves up and down according to temperature, water level, and other factors. The workers may also move the queen deeper in response to disturbance.

The queen never goes out. Indeed, only the eldest and most expendable 10 to 20 percent of her offspring leave the nest to forage for food. They exit not through the mound, but through horizontal tunnels radiating out as much as 130 feet, and, once on the surface, they cover the territory with a looping pattern that can be dismayingly efficient. (In Texas, they frequently get to highway accident victims before the ambulance.) If they find a substantial food source, they make a beeline back to the nest, navigating off the sun or moon and laying down a chemical trail for other foragers to follow. They share the food with the homebodies in the nest by regurgitating it so it can be licked or sucked from their mouthparts. The recipients in turn regurgitate part of the meal for their unfed cohorts until everyone is equally full, or empty. It may sound disgusting and also suspiciously communistic, but it makes for quick, even distribution of food through the nest, while at the same time buffering the queen and her brood with two or three tasters from any dietary indiscretion. To get to the queen in the first place, an effective fire ant poison thus needs to have a delayed action, over three days or more. It must be short-lived, for environmental reasons. And it should kill only imported fire ants.

In the aftermath of mirex, biological control seemed like the most promising area for new research. One explanation for the imported fire ant's extraordinary success in this country was the phenomenon of "biological release" from the checks and balances of its native habitat. Like any traveler in a new country, the fire ant was suddenly free to eat, drink, make love, be merry. It had neither friends nor enemies. USDA researchers in Gainesville began looking up its old acquaintances in the hope of introducing them into this country to restore control, and in the past decade, they have identified more than a dozen possibilities in the fire ant's native range, a fifteen-hundred-mile strip running from Rio de Janeiro up into central Brazil. Unfortunately, none of them now appears likely to work in the United States. Some fungi and bacteria are too big to get down the fire ant's throat, or they produce no ill effect, or if they do, the fire ants heave the sick and wounded out of the nest to avoid contagion. A parasitic nematode sounded promising for a while, but it proved necessary, as one researcher put it, to "practically drown the ants" in nematodes to get adequate penetration. Fire ant colonies took exception to being inundated in nematodes. They tended to pick up and move, leaving disconsolate researchers behind.

Another approach to the perfect fire ant poison is the use of the ant's own pheromones. These chemical messages are by definition attractive only to the species that produces them. In Gainesville, a slight, scholarly chemical ecologist named Bob Vander Meer has analyzed the pheromones of *Solenopsis invicta* with gas chromatography and reproduced them artificially. On the wall of his laboratory is a photograph of two patches of paper. One, impregnated with the fire ant's trail pheromone, swarms with foragers; the other, untreated, is almost empty. Vander Meer aims to combine the trail pheromone with existing poison baits, so that only imported fire ants pick them up. But he doesn't know yet if it will be possible to synthesize the pheromone in quantity at a reasonable cost.

In short, there is no magic bullet. A visitor to the USDA's Gainesville fire ant center or, in truth, to almost any other fire ant research program in the country, comes away with a sense of frustration and futility. But if the researchers gripe about the politicians, their budgets, and even each other (nobody seems to think the other guy's approach has a hope in hell), they share a grudging admiration for the fire ant itself, which has been outsmarting them for forty years. "It's like a wife," said one scientist, best left unnamed. "You think you got her all figured out and then she turns around and does something weird on you." The implication is that people are going to have to learn to live with the fire ant, which begets the question of just how bad that really is.

I live in Connecticut, where we don't have firsthand knowledge of the fire ant, and a lot of friendly southerners suggested it might be an educational experience for me to apply some vital part of my anatomy to a fire ant mound and just see. (Some researchers told me that a hybridized fire ant more tolerant of cold could ultimately range up the East Coast to Long Island, and up the West Coast into Washington, a development for which I was content to wait. But others dismissed such predictions as alarmist nonsense intended to shake loose more federal research dollars.) In the end, I decided not to stick my hand in a fire ant mound, on the theory that a working journalist should strive to be no more than half as dumb as a politician seeking office.*

What I did instead was put my hand and some fire ants under a microscope. Magnified sixty-three times, a fire ant looks monstrous and hairy. When I dropped one accidentally, I leaped back from the microscope in alarm. I picked it up again, somewhat abashed, with a pair of tweezers. At the end of its glossy, segmented body, the uplifted gaster twitched angrily. (The fire ant appears to spray its venom as a warning or repellent to insect enemies. In a case of what Vander Meer calls "functional parsimony," the venom may also serve as an antifungal fumigant in the nest.) The fire ant's head was broad and flat, with poppy-seed eyes at the sides. Its antennae paddled the air, and two serrated mandibles, each like the thicker half of a lobster's claw, tried to grab hold of the tweezers and fling them up into my eyes.

At the same magnification, the back of my hand looked like Jell-O shrink-wrapped in polyethylene. Here and there, black hairs stuck like arrow shafts. Fire ants roamed through this

---

*I guess I should say a *print* journalist because soon after I wrote this in the pages of a national magazine, I became involved in television, and almost the first thing I was obliged to do was kneel in front of a camera while sticking my hand in a fire ant mound and saying "Hello, my name is Richard Conniff." We did about twenty takes because, after the first few dozen stings, I had a little trouble pronouncing my name and eating my words at the same time. Also the camerawoman kept talking about how beautifully the light fell on my contorted features. For the next week or so, my hand was a mass of welts and pustules.

wilderness, and having determined by subtle chemical analysis that I wasn't a nestmate, stung me at will. The only trouble was finding them through the microscope and keeping my hand still to stay in focus. Finally, though, one wandered into view and I steeled my resolve. It grabbed hold of some skin with its mandibles, tucked its hind end under like a dog about to do its business, and zapped me. Then it pulled out the stinger, which is a smooth lancet, not serrated like the honeybee's, moved one step to the side, and started to do it again, at which point I moved.

It was worse than a mosquito bite but nowhere near as bad as a wasp sting. On a scale of relative sting pain prepared by a welt-covered USDA scientist, a Central and South American ant known as "the bullet" (because you feel as if you've been shot) rates a four, the paper wasp a three, the fire ant, despite its name, a mere one or two. This is far milder than even one of our own homegrown species, the harvester ant (a three), with which the Apaches sometimes amused their prisoners. Does this mean the fire ant peril is a canard?

Unfortunately, no. The rating scheme doesn't take quantity into account, and quantity is the fire ant's great strength. It is probably less common to get one fire ant sting than ten or twenty, and even if the pain passes quickly, a splotch of itchy yellow pustules usually results. They can persist for a week or more on a child's arm—an ugly memento mori to anxious parents—and though the pustules are sterile, scratching often leads to secondary infections.

Quantity, even superabundance, is characteristic not just of the fire ant's sting, but of all its behaviors. With cockroaches, people see one or two at a time, and after a few days they may call an exterminator. "With a fire ant," said one researcher, "you see four hundred of 'em lining up and coming across the table, and you call the pest control man *now*." Superabundance is particularly an issue in Texas, where the fire ant is once again doing something weird—even potentially horrific—causing one state magazine to brand it "the ant from hell."

In 1973, a USDA researcher named Mike Glancey was digging up some fire ant nests in Hurley, Mississippi, when he came across something no one had seen in thousands of similar excavations: Instead of one queen, there were "twenty or thirty big fat girls without wings that looked like queens," and instead of executing these pretenders to the throne, "the ants were tending them, surrounding them, protecting them." He dubbed the nest "queen city" and wrote it up. The scientific community responded with disbelief until 1980, when reports of other multiple-queen colonies began to come in, particularly from Texas. Some colonies had three hundred queens, all laying eggs, and the colonies were packed together in unheard-of densities. Thus far, multiple-queen colonies have appeared only in pockets around the South, but in one extreme case, parts of a field were infested at a rate of 22.6 million fire ants per acre—519 individuals per square foot.

In truth, quantity has always been the key to the imported fire ant's notoriety. North America's native fire ants, now largely forgotten, could also sting. But native fire ants typically occur at densities of only four mounds per acre. In its single-queen version, the imported fire ant builds ten times as many, an increase horrifying enough to provoke a twenty-four-year pesticide war. Hence the alarmed reaction to multiple-queen colonies, which represent yet another tenfold increase.

Texas is the only state where multiple-queen colonies predominate. The mounds there are sometimes so close together, according to the USDA's Sanford Porter, that if you're light-footed and dim-witted enough, "you can practically hopscotch across your yard on fire ant mounds like a frog on lily pads." Not only are the colonies close together, but they also lack the old territoriality. They appear to share food and they no longer automatically massacre any intruder not chemically labeled as a nestmate. Dump one colony onto another and they will probably send out for beer. The mounds function, in effect, as a single supercolony over an acre or more in size.

Scientists at first thought the advent of the multiple-queen form might represent an overnight case of "evolution under our noses," or perhaps the introduction of a new imported fire ant not yet properly recognized as a separate species. But genetic analysis has since demonstrated that multiple- and single-queen forms are both *Solenopsis invicta*. Researchers in South America have learned that fire ants sometimes display the multiple-queen form in their native habitat. The two forms may simply be alternative ways of populating the landscape: When the disturbed areas that fire ants like are patchy and widely distributed, the single queen's nuptial flight may be the most efficient way to find them. But when large contiguous areas of disturbed land are available, as on a Texas ranch, it might make more sense to build up a huge multiple-queen (or polygyne) colony, which will ultimately split and disperse by simply moving next door.

The multiple-queen fire ant can have a startling effect on its habitat. In one study, Sanford Porter timed how long it took fire ants in the wild to locate baits he set out. In single-queen areas, the first foragers showed up in two minutes. In multiple-queen areas it took just twenty seconds, a rate of efficiency with dismaying implications for other animals in the habitat. Porter, who wears thick, silver-rimmed eyeglasses and a trim moustache, is a careful scientist prone to laconic wit, but the title of his study report reads like a tabloid headline: "INVASION OF POLYGYNE FIRE ANTS DECIMATES NATIVE ANTS AND DISRUPTS ARTHROPOD COMMUNITY." When multiple-queen fire ants entered his study plots, he found that the number of individual ants increased as much as thirtyfold—almost all of them fire ants. The number of native ants dropped 90 percent, and the number of native ant species dwindled from fifteen to five. It was the first solid evidence for a troubling impression a lot of Texans have been getting lately—that multiple-queen fire ants may be simplifying the environment, stripping away biodiversity, setting the countryside up for another silent spring, this time without benefit of pesticides. "Texas," wrote Porter, "may be in the midst of a genuine biological revolution."

Anecdotal evidence abounds. Bird-watchers have discovered fire ants climbing up poles and into nests to eat bluebird hatchlings. Small-mammal researchers have learned that fire ants will strip any animal left overnight in a live trap. Larger animals may also be vulnerable. After one state park treated a severe fire ant problem, the reptile population rebounded so well that it was necessary to reinstall a "Beware of Snakes" sign.

The trouble with anecdotal evidence is that one dead bluebird hatchling or even a hundred dead fawns don't necessarily add up to a species in trouble. One of the original arguments for using heptachlor in the 1950s was that imported fire ants were wiping out the quail and bobwhite. It developed afterward that fire ants, unlike heptachlor, mainly killed unhealthy chicks, with no significant effect on overall populations. They were just another one of nature's gruesome mortality factors.

Thus the fire ant, a creature with a submicron-sized brain, has Texas on the horns of a dilemma: Does it make sense to treat critical wildlife habitat broadly and aggressively, risking the ill effects inherent in any pesticide campaign? Or is it wiser to wait for conclusive evidence of a new silent spring, at which point it may be too late?

In a recent case rich with irony, the Audubon Society, long-standing opponent of irresponsible pesticide use, brought its fire ant problem to the Texas Department of Agriculture. The department has become known for its enlightened approach; it speaks not of eradicating fire ant populations, but of *managing* them. In practice, this means persuading people that if they're lucky enough to have native fire ants they should treasure them as a stopgap against imported species. It means telling angry farmers and homeowners that they can live with forty mounds to the acre, and if they choose to kill their single-queen colonies, they'll probably wind up with multiple-queen super-colonies instead. It means getting people to forgo the urge to kick, cuss, and kill, in favor of slower-acting but more effective remedies—notably the use of a growth regulator that renders queens incapable of egg production, but may take as long as a year to shut down the colony. It is a formidable chore because people have always overreacted to fire ants.

The Audubon Society was understandably wrought up because imported fire ants on an island refuge near Corpus Christi were killing brown pelican hatchlings, a threatened species. Caretakers could actually watch it happening, and death by fire ant is appalling even when the victim is a lowly mealworm. In some areas, more than half the hatchlings died. The department of agriculture sent out a team to do a survey, which found imported fire ant mounds all around the nesting area. It recommended treating for the imported fire ants, but not for some native fire ants minding their business off to the side. "You don't understand," the Audubon man replied. "We want the ants *off the island.*"

It is easy to see in this the hand of the fire ant as cunning old joker, amusing itself at human expense yet again, putting the Audubon Society in roughly the position of an antipornography preacher showing up at a peep show and demanding to see the hot stuff—another of the fire ant's characteristically masterful ironies. In the end, the enlightened approach prevailed: The department of agriculture man treated the imported fire ants and reported that a rattlesnake kept him from covering the rest of the island.

But is it any wonder that fire ants drive people to raving vexation? Is it any wonder that the sight of them advancing en masse down a wall, or tumbling out of an electric switch onto a child's hand, or blinding a fawn with stings should cause even environmentally enlightened sorts to wake up from their seething nightmares yelling, "Quick, Henry, the Flit"? "Hell," said one researcher, "if you told a guy that plutonium could kill fire ants, he'd want to know where to buy some."

The trouble is, it would be just like the fire ant to survive.

The Big Calamari

Roaring seaward off the coast of Florida, the boat, a rehabilitated drug runner, sluices up veils of spume. Water washes in sheets across the windshield. Rainbows flash and atomize at the rails. On a captain's chair amidships, Clyde Roper sits, contented, earplugs in, bracing himself with feet set wide.

He is an unusual character, especially considering the ungodly quest he has undertaken. It's as if somebody crossed Ahab with everybody's favorite high school science teacher. He is fifty-eight years old, with a gray jawline beard and a New England lobster-fisherman's accent, earned honestly, as a teenager working his own boat in New Hampshire. He limps slightly from an Achilles tendon ruptured, he confides, while "parasailing for giant squid."

Actually, he fell off a ladder, but Roper is not above a little blatherskite in the cause of science. He once arrived at a lecture bearing a four-foot-long squid on a hospital stretcher and, taking the podium, inquired, "Okay, who ordered the calamari?"

Today, Roper will be collecting larval squid on the edge of the Gulf Stream. The boat, operated by the Smithsonian Marine Station in Fort Pierce, stops seventeen miles out, in 750 feet of water. Roper bolts a yellow sledge to the trawl cable and sends it over the side, yelling, "Clunk in," to the man at the big blue Sea Mac winch. The plankton net follows, a ten-foot-long cone of fine mesh tapering down to what looks like a mayonnaise jar. It billows out, gleaming a ghostly bluish white beneath the surface, then sinks out of sight. The catch will consist of planktonic life-forms visible mainly with the help of a microscope. But Roper comes to the chore in his characteristic spirit of heroic optimism.

Roper, a zoologist at the National Museum of Natural History, has sailed over every ocean on Earth, and under many of them, studying squid, octopus, and cuttlefish, together known as cephalopods. He has studied how some cephalopods use light to communicate in the rich language of bioluminescence. He has described minute structures, like goose bumps, in cephalopod skin and theorized about what role they may play in hydrodynamic streamlining. He has also identified one new family, two genera, and about twenty species. As the culmination of his life's work, he is preparing to lead an expedition in search of one of the last great monsters of the sea, the giant squid, described by Herman Melville himself as the "white ghost . . . which . . . few whale-ships ever beheld, and returned to their ports to tell of it."

Architeuthis, the giant squid, is the largest invertebrate on Earth, and yet even now, despite sonar and submersible vehicles, no one has ever seen it in its natural setting, five hundred fathoms down. We tend to think of invertebrates as puny, mindless things like midges and slugs. But this is as misguided as if we formed our view of vertebrates based only on minnows, without ever having seen an elephant. *Architeuthis* can grow to a

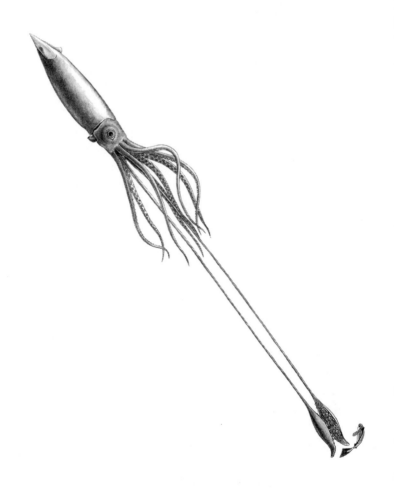

length of sixty feet, half again as long as a school bus. It has eyes the size of headlights. In addition to its eight arms, it has two lengthy tentacles ending in sucker-lined clubs, which it can lash out to seize its prey. Nestled amid the arms, its parrotlike beak is powerful enough to crush bone.

What we know about the giant squid is largely superstition, equivalent to what a medieval bestiary writer who had seen ivory tusks, and heard that elephants copulate back-to-back (out of modesty), could tell us about the actual animal living somewhere in the impenetrable depths of Africa. Monstrous cephalopods abound in our mythology, beasts like the many-armed Scylla in Homer's *Odyssey,* and the multiheaded Hydra destroyed by Hercules. In *20,000 Leagues Under the Sea,* which became a 1954 Walt Disney film, a giant squid grappled ferociously with the submarine *Nautilus.* Jules Verne supposedly based this epic battle on an actual encounter between a French warship and "a colossal and slimy embryo" that could open its parrot-beak mouth "about one-and-a-half feet." Not to be outdone in the realm of sea monsters, Peter Benchley depicted the giant squid, in his 1991 novel *Beast,* as a man-eater driven to "instantaneous explosive spasms of destruction." It was also prone to clambering up over the transom of large fishing boats (about the size, say, of the one in which Roper is now traveling), snaking its horrible tentacles across the deck, and dragging unsuspecting crew members to their deaths. Benchley's squid actually pops open a submersible like a can of nuts and gobbles up the three passengers. In roughly the same vein, Roper keeps a Gary Larson "Far Side" cartoon taped to a wall in his office in Washington, D.C. It features a pet giant squid that ate the neighbor's Shi Tzu.

One difference between us and the medieval bestiarists is that *our* superstitions are about an animal found on our own coast. By Roper's calculations, in fact, giant squid ought to be commonplace. They appear to inhabit almost every ocean, and roughly two hundred of them have turned up over the past century stranded on beaches. Roper himself collected the battered carcass of one such giant squid in Massachusetts in 1980; it's now on display in the museum, preserved in a clear acrylic coffin full of isopropyl alcohol, which Roper calls "the largest martini in Washington, D.C." Giant squid are also common enough that, in thirty separate incidents over the past ten years, they came to the surface tangled in deep-water fishing nets. Still, no one knows how they live in the absolute darkness of their own world. We cannot even say if there are three species of *Architeuthis,* as Roper thinks, or nineteen species, as other researchers have suggested.

After fifteen minutes, the man at the winch yells, "Surface!" and the first tow comes in. The plankton net is flocked with pink crud, and the mayonnaise jar overflows with goo. Roper summons up his reservoirs of scholarly knowledge and concludes, "That's what goes under the category of snot," spilling out the sorry remains of a jellyfish. He hauls the net up on a davit, hoses it clean, and sets it out for the second tow, utterly undaunted.

In between, the battle plan for his expedition emerges, and it appears to hinge on finding "a friendly sperm whale" and letting it lead him to the squid. Sperm whales, he explains, routinely prey on *Architeuthis.* Ambergris, the precious waxy substance used in perfumes, is actually a product of the whale's intestines designed to ease the passage of indigestible squid beaks. Roper estimates that a fifty-ton whale may eat three or four large squid of various species every day. It catches the squid in its mouth and will occasionally wrestle it, still struggling, to the surface. The whales almost always bear scars of the squid's sharp-toothed suckers around their mouths.

The notion of a "friendly" sperm whale might give a prudent man pause, but Roper is patient and methodical. "I've been thinking about this giant squid business for a long time," he says; in fact, it has been thirty years. One day, when Roper was a graduate student in the School of Marine Sciences at the University of Miami, his department got a report of a sperm whale stranded on a nearby beach.

"We went for our dissecting kits, which consisted of double-bitted axes and a couple of hunting knives. As we left the lab, we grabbed a book off the shelf because, as invertebrate zoologists, we didn't know Shinola from beans about whales."

("Shinola from beans" is not a bowdlerized quote, by the way; Roper, who is the son and husband of schoolteachers, really talks that way. He is an impeccably nice man. Blatherskite, yes; billingsgate, no.) Roper ended up inside the belly of the whale. "She was still warm," he recalls, and then the optimist in him asserts itself. "But if you're going to work on a dead whale, it might as well be fresh. I got back inside the third stomach and, sure enough, I could reach way down and bring out handfuls of squid beaks."

Roper emerged saturated in the whale's black blood and under assault by every biting insect in Florida. He went to rinse off in the surf, but a shark had picked up the whale's scent. So he and his gore-smeared colleagues hopped back in their Rambler wagon, with the whale's jaw strapped on top, and headed home, inflicting olfactory trauma at every tollbooth on the Florida Turnpike.

"I've looked at the biology of the giant squid," Roper continues, "and also of the sperm whale, the records of where strandings occur, and the best concentrations of squid in stomach contents of sperm whales." Gradually, he narrowed the search down to an area off the coast of New Zealand. Last year he made scouting trips there and to the Azores, another hot spot for both sperm whales and giant squid. In the Azores, he interviewed whalers who had worked in the trade as recently as 1981. "It turns out that when a harpooned sperm whale comes to the surface, just before it dies, it vomits. So I said, 'That's great, what's in the vomit?' and they looked at me like, 'Man, is this guy weird,' and then they said, '*lula*,' which means squid. What kind? They said '*lula grande*.' How *grande*? And every one of them would take both hands and grab their thighs, meaning the arms and tentacles were that big, and these guys were not spindly-legged wimps. I asked how often they ran into this, and virtually every whale would have a giant squid in it.

You do a little arithmetic—there are 400,000 sperm whales in the world—and there have to be a lot of giant squid around."

The second tow comes in, and Roper digs into it with relish. A couple of assistants help him wash down the contents of the plankton net. Then all of them hunch low, as if shortsighted, over the tray into which they have transferred the concentrated catch. "Ooh, what a lovely soup!" Roper exults. In truth, it resembles cold miso, cloudy and brownish, and they fish through it till their bare fingers turn pruney. "That's a squid right there. Fantastic! Outstanding!" Roper says, indicating a squiggle. "They're both squid. No, one is a little fish." Actually, it's one of Florida's great game fish, a blue marlin, also larval, small enough to lay across the pad of his fingertip. Roper points out a squid named *Abraliopsis,* and you can just make out the two black photophores on its arms, which can produce a flashing light to scare off predators. "Oooh," says Roper, "Alright! Gorgeous!"

It is a soup within a soup: Beneath the boat, the Gulf Stream churns steadily to the north, bearing a rich stew of larval squid and other life-forms to supply the fisheries from the Middle Atlantic States up to the Maritime Provinces. There may be giant squid of almost any size out here, under the keel, even now; like other cephalopods, they start out small, and may blossom to adult size in just three years.

The ocean surface is flecked with lazy little whitecaps, building toward a boil. A hurricane is looming up, piling dark clouds overhead. As the boat turns back for port, Roper sits down, with the day's entire catch preserved in a few small vials in his shirt pocket, and attempts to explain his cephalopod obsession.

The biological class known as Cephalopoda, meaning "head-footed animals," consists of two subclasses. On the one hand there are the half-dozen chambered nautilus species, a family of evolutionary throwbacks with an external shell. The nautilus has remained largely unchanged for 450 million years. But drifting along like a snail, snug in a shell, ultimately failed most other cephalopods, Roper says. Large, predatory fish evolved to swallow them whole, digest out the meat, and excrete the shell. Hence the second subclass, the Coleoidea, consisting of squid, cuttlefish, and octopus.

Natural selection turned these other cephalopods inside out: The flesh gradually engulfed the shell and reshaped it into an internal stiffening structure. In squid, this structure is a long, flexible rod called the gladius, a word Roper explains with a snicker-snack sword-fighting flick of the wrist. In cuttlefish, it's a flat, oblong disk, which has become a familiar fixture in bird cages, to be pecked at as a calcium supplement. Fossilized cuttlebone is also the raw material for the carving of meerschaum pipes.

By internalizing their shells, the cephalopods accomplished an extraordinary evolutionary makeover: Once dull as clams, they became svelte and slinky free-swimmers. They also evolved a highly complex way of life, characterized by speed, stealth, and intelligence.

"You've heard the phrase 'the observed observer?' " Roper asks, recalling what it's like to swim in small schools of foot-long Caribbean reef squid. "You know they know you're there. They will always, no matter how careful you are, maintain what is for them a critical distance. They won't be bothered by you, but if you swim forward into the critical space, the squid will move away. If you stop, they'll stop. If you back up, they'll move toward you. They have these great big shiny eyes, just gorgeous. They're really tuned in to what's going on around them."

The cephalopods have evolved the largest brains of any invertebrate, which might sound at first like damning with faint praise. But their brains are generally bigger, in proportion to their body size, than those of their vertebrate competitors, the fish. What they do with them remains largely a mystery. In a Peter Benchley thriller, anything as big and brainy as a giant squid naturally must be a monster of "perverse malevolence."

But cephalopods lead far more sophisticated lives than we might suppose. Octopus survive well in captivity and perform impressively on intelligence tests, learning and remembering simple tasks, like how to negotiate a maze; one octopus may actually teach another. Cuttlefish seem to recognize their individual caretakers. But most squid, built for open water, are too delicate to survive in captivity. We know more about how they behave on the dinner plate, *fra diavolo,* than in their own world.

Even the basic anatomy of the cephalopods—simple stuff like telling head from tail—can seem bewildering to the untrained human eye. It does not help that they can swim backward or forward at will. The usual cephalopod arrangement is to have the parrot-beak mouth in the middle, with the arms and tentacles arrayed around it like petals on a flower. The arms pass food into the mouth, where a tongue lined with raked-back teeth rams it down the gullet. Just behind the mouth, the esophagus passes through the middle of the brain. So if a cephalopod takes too big a bite, Roper notes, "it'll ream its brain out." (This may be a source of comfort to nightmare-prone mariners: Even a giant squid is unlikely to swallow a person whole. It would surely chunk you up into little bites first.) The outsize eyes are just behind the mouth. Then comes the cigar-shaped mantle, which often ends in broad stabilizing fins at the tail.

But within this basic design, the cephalopods are amazingly diverse. Some cephalopods press their arms together and shoot tail-first through the water sleek as torpedoes. Others drift upside down along the bottom, with their tentacles sprawled out, trolling their sticky suckers through the planktonic soup. When threatened, many deep-sea squid wrap their arms back over the head like a skirt, exposing rows of suckers armed with sharp teeth. One bright-red squid leaves only its squared-off tail exposed, and it's guarded by what looks like a diabolical imitation of a face. Around a puckered spot like the mouth of a dyspeptic old witch, two false eyes glint with bioluminescent menace. A crescent-shaped "eyelid" actually seems to blink across their cold surface. Scientists have named the species *Vampyroteuthis infernalis,* the vampire squid from hell.

With cephalopods, it's hard to resist the urge to become fanciful. Roper and a colleague once named a whole family of squid Batoteuthis, from the Greek "batos" meaning "thornbush," ostensibly because the squid has a long thornlike tail and large bushy clubs. But this was also the middle of the Batman craze, and if you look closely at the illustrations accompanying Roper's article in a learned oceanographic journal, you may notice a tiny bat of unknown species flying out of the squid's oviduct. (Roper protests that he thought the editors would white it out.)

But the fanciful appearance of the cephalopods is nothing compared to their behavior. Intelligence in this world may have less to do with mazes than with the subtle language of light and color. "In the skin are gezillions of chromatophores, little sacs of pigment," Roper says, "each of which is individually controlled by nerves that come from the brain. Cephalopods can mix and match colors for a number of functions." They can change their color, shape, and position in the water to blend in with a spiky bed of coral. When swimming in the deep sea, squid may hide themselves with complex displays of light on their undersides. Roper recorded one squid as it varied the intensity of its bioluminescence 16,000-fold, in perfect synchrony with the shifting patterns of downwelling sunlight.

Cephalopods are famous for escaping behind a cloud of ink. The cloud is actually a mix of ink and mucus, which does not conceal, but hangs together in the shape of the cephalopod, as a decoy to distract predators. In some deep-water species, the so-called pseudomorph, or "fake body," is actually bioluminescent, as if to say, "Bite here."

Shallow water cephalopods may also put on a show of color to mimic some poisonous or predatory fish, or as a warning during courtship. "Sometimes a male will lie side by side with the female," says Roper, "and the side that's away from the female will be lit up with this incredible flame-stripe pattern." He draws his arm across his face like a cape and peers out over his elbow: "*Flash, flash, flash!* 'Stay away from her! She's mine!' "

In courtship, male and female may swim side by side for hours, shifting direction together as smoothly as Fred Astaire and Ginger Rogers in heaven. They swim backward or forward with equal grace. If they switch sides, the male instantaneously adjusts his display, to keep the warm seductive tones toward her, and the flashing neon "drop dead" signal toward interloping males.

Part of the secret of their success is jet propulsion. The cephalopod mantle draws water in through openings behind the eyes and shoots it out again through a funnel, located just under the head. The squid can swing this funnel around like a firehose to point in any direction.

To master jet propulsion squid have evolved the largest single nerve fiber in the animal kingdom, a hundred times thicker than a typical human nerve fiber. Neurobiologists have been studying this so-called giant axon for more than fifty years, and much of what we know about how nerve cells generate and conduct electrical messages comes directly from squid. But Roper would rather talk about what the giant axon does for the squid itself: "The squid sucks water into its mantle. Then it locks the mantle orifice with two little lock-and-key mechanisms. So the signal comes from the brain to all those rings of muscle in the mantle and says 'Contract.' " If the signal caused the rings nearest the head end to contract first, the water would be squeezed back away from the funnel and into the tail. "The squid would essentially blow its rear end off."

The giant axon prevents this catastrophe, because of one of the most basic rules of neurobiology: The speed of a nerve signal is directly proportional to the diameter of the nerve fiber. So the signal flashes instantly down the giant axon to the tail, then out the branching nerve fibers. Everything is engineered to the exact length and diameter so that the signal reaches the hindmost ring of muscles first. Then the next-closest ring of muscles contracts, and so on sequentially to the head. Instead of exploding squid parts, the result is rhythm and harmony.

Roper's obsession with cephalopods takes one other form, and sensitive readers may just want to put this book down right now, and instead blissfully daydream of swimming in their Speedos among the cephalopods, to the sort of shimmering-jellyfish music found in television shows about the jewels of the deep. For readers of sterner appetite, what follows will assuredly lead us back to the subject at hand, which (in case you've forgotten) is *Architeuthis,* the big calamari.

Roper is passionate not just about cephalopods, but also about growing Thai hot peppers, jalapeños, and habañeros, and the two passions often come together in a frying pan. Having delivered an enthusiastic description of the differences among various cephalopods, he is liable to conclude, "Which one tastes best depends on how much garlic you use." In the cephalopod room at the museum, he keeps thousands of specimens in jars of preservative—and also an aromatic string of dried octopus and cuttlefish from the Hong Kong market, which he sometimes sniffs with great gusto.

"Cephalopods are about the perfect food," he says. "If you eat the mantle, there is virtually zero fat. The only way to get fat is to eat the liver, which they do in Japan. I must say, it's an acquired taste, and the speed of acquisition is directly proportional to the amount of *sake* you wash it down with." Even for more conventional appetites, only about a quarter of the cephalopod goes to waste, versus about half of most fish.

Cephalopods are short-lived, fast-growing, prolific animals, all the things a robust fishery requires. Driftnet fishing for squid has recently been banned, not because the squid populations were threatened, but because squid often served as a cover for illegally catching more valuable fish, like salmon, and because the nets incidentally killed thousands of sharks, dolphins, and other species. But the worldwide catch is still more than two million metric tons a year, and cephalopods are increasingly common even in the squeamish American market.

Roper, who sometimes teaches classes in cephalopod cookery, recommends this squid recipe for diners seeking enlightenment: Briefly stir-fry a mixture of the chopped arms and tentacles, garlic, uncooked rice, mint, raisins, and pine nuts. Clean out the mantles and stuff them with this blend, adding a little red wine. Sew shut. Add olive oil and then tomato juice diluted with clam juice to cover. Bake at 350 degrees for 45 minutes, until tender. Baste cook as needed with *sake*. After that, you may be ready for octopus salad.

There is a venerable history of cephalopod scientists eating their research material. Over fried calamari one night, Roper recalls the eccentric nineteenth-century naturalist Constantine Rafinesque, who taught at Transylvania University in Kentucky, where Roper himself was later an undergraduate. Rafinesque once reported a new species of octopus and wrote, "I did eat this *Ocythoe* which afforded a meal for many, and it was as good as usual with the Octopus." Unfortunately, what he had eaten was the only known specimen, and *Ocythoe tuberculata* was not found again for another eighty-two years.

But Roper's culinary adventures have yielded happier results. One time, Roper was serving as an examiner for a doctoral candidate in Newfoundland. "It was a very long British-style examination, in academic robes." Afterward, the new Ph.D. and his wife invited everyone home for dinner, which consisted of cephalopods in twelve different dishes, plus beer and a local rum called screech. "So we got to talking about giant squid," Roper says. "I wondered if giant squid would be any good to eat, because, what a great way to get a lot of food, and the new Ph.D. said, 'Well, I have a piece in my freezer.' He was a cephalopod man, and a lot of giant squid strand in Newfoundland. So we thawed it out and prepared it, and it was awful, very bitter. One of the other examiners and I looked at each other and at the same time we said, 'Ammoniacal!' "

This was literally food for thought. One of the questions about the giant squid is whether it is a fast, powerful swimmer, as implied by what could be termed the monster hypothesis. But the ammonia in its muscles suggested otherwise. Ammonium ions are lighter than water. They should enable the giant squid to maintain its buoyancy without spending a lot of energy swimming. Following that dinner, Roper and his colleagues determined that *Architeuthis* muscles are soft and spongy, and that their jet propulsion equipment is less powerful than that of some smaller squid that are known as strong swimmers. Hence the sluggard hypothesis: Giant squid are basically big, lazy cruisers, because they can afford to be. They can get enough to eat by simply hanging around and, every now and then, shooting out their tentacles like a frog's tongue to snag some passing fish. No speed known to squid, Roper believes, would save them from their one predator, the friendly sperm whale.

With all this in mind, Roper's plan is to find his whale, probably resting at the surface somewhere off the coast of New Zealand. He'll attach a television camera to its back and, vicariously at first, ride down with it when it dives. If his luck borders on the miraculous, the television camera might actually give him a whale's-eye view of a giant squid in the moment before it gets eaten. But it's far more likely, at the depths where whale and squid meet, that he will see only a great deal of darkness. So Roper will also track the whale acoustically. "Some researchers think that they hunt by sonar. The pattern of the sonar signal starts with a slow clicking, and then it becomes increasingly rapid, until it's almost a buzz. Click . . . click . . . click-clickclick . . . zzzhh . . . and then it stops. *Boom!*" He slaps his hand down on the table. "And that's where the animal is captured." It may turn out, however, that sperm whales don't hunt by sonar at all, but by vision.

By working with the whales over the course of his four-month expedition, Roper believes they will lead him to where the hunting ground for giant squid is richest. "So much of this is about looking in the right place," Roper says. "You don't go to a frog pond to catch trout."

From the mother ship, Roper will be able to send down unmanned submersibles equipped with powerful lights and cameras. He will also descend himself in the bubble-fronted Johnson Sea-Link submersible. "I see us moving along very slowly and as quietly as we can in the sub," Roper says, "and seeing first in the distance as far as our lights can go, a hazy form, and suddenly realizing that it's an *Architeuthis* out there. I'm hoping that it will not be frightened, so it allows us to get close enough to see the whole animal. We'll be able to learn in a few seconds more about the animal than we have ever known before.

"We'll see how the animal orients itself in the deep sea, horizontal or vertical, head up or head down. When we look at the giant squid for the first time, we will be able to tell if it uses its fins for swimming. Does it hover? Does it sit on the bottom? We may be able to determine how it holds its tentacles, which are extraordinarily long. If we are exceptionally fortunate, we might be able to see it stalk and capture prey."

The conversation wanders to the question of whether the giant squid might produce a pseudomorph, and whether it would be bioluminescent, or stretch out as long as the animal itself. Caught up in all the possibilities, Roper says, "It would be wonderful to get the giant squid to think we're a predator, and exhibit the full range of defensive behaviors. . . ." He hesitates a moment.

"The other thing that would be wonderful . . ." Roper's eyes light up behind his bifocals and he cackles with anticipatory glee ". . . would be to have the giant squid attack us.

"Can you picture us being in this big fish bowl, five or six feet in diameter, and having this giant squid completely envelop it, with the suckers all around? And seeing the beak just this far away, scraping against the acrylic? Wouldn't it be great to see?"

A companion nods with widened eyes, a little dumbfounded. The bowls of calamari are empty now, and somewhere between eating squid and being eaten by them, it seems like a judicious moment to say goodbye and leave Roper to his quest. "And it would be time to surface at that point," he concludes. "Blow all the ballast, but keep the camera running, and radio up to say, 'Hey, we have a friend on board.' "

Lions of the Pond

A hot summer morning at the Slaughter Ranch in southeastern Arizona, everything lazy and slow. The pond by the house is stippled with small green knobs that look like waterlily buds, but turn out, on closer examination, to be the golden eyes of a hundred hungry bullfrogs. The surface is clotted with mats of yellow algae. There is no wind to stir the willow trees. Only the dragonflies are moving. They thread the air in a thousand directions at once. Red ones that are as fine and slender as sewing needles. Big ones, like winged snakes, with jade and turquoise patches enameled on their long abdomens.

"There's *onusta* coming around that tree," a dragonfly enthusiast calls out, in a nasal voice. "No, that's *saturata,* with *multicolor* just out ahead."

"You sound like you're calling a horse race," another collector says.

"And it's *saturata* by a head!"

"By a frons!"

But a racetrack isn't quite the right metaphor. Too many other things are happening here. Territories are being gained and lost, rivals run off, mates wooed. A happy couple darts through in the wheel position, the male flying for them both, the female hanging on underneath like a whooping thrillseeker on an amusement park ride.

A member of the species *Anax junius,* built like a 747, patrols overhead, then hesitates in midair to seize a fly and gnash it with its bomb-bay mouth. The racetrack announcer coils and swings, his net whistling through the air. He extricates his catch, its huge wings held together between thumb and forefinger, and presents it to the youngest member of the Dragonfly Society of the Americas, an eight-year-old in a web vest and round, wire-rimmed glasses, named Ben Lane. It's a brilliant blue-and-green creature, the length and breadth of the boy's hand. Its eyes are opalescent pools. The racetrack announcer explains that *Anax junius* means "Lord and Master of June" and the hunting behavior is called hawking.

It becomes clear, the longer Ben holds the dragonfly without being eaten alive or lifted into the treetops, that the most dangerous thing about dragonflies is the danger of mixed metaphors. Dragons, horse races, and hawks are the least of it. Because they're big and easy to see, dragonflies have attracted a richly varied, mostly unflattering mythology.

One branch of folklore holds that dragonflies are servants of snakes and revive them, Lazarus-like, from the dead. Hence the southern name "snake doctor." Dragonflies are also "devil's darning needles," capable of stitching together the lips of wicked children in their sleep. They are "eye stickers" or "mule killers." Given the nasty reputation of insects in general, it probably figures that a syringe-shaped insect with a five-inch wingspan and eyes lit up like Christmas ornaments might seem vaguely menacing.

But we know better now. Dragonflies don't sting. The deadly-looking hind end is a kind of tow truck grappling hook in the male, and, in the female, an ovipositor, or egg-laying device, not a stinger. They don't bite either. Dragonflies (and their close kin, the damselflies) are members of the order Odonata, meaning "toothed ones." But humans who are uncommonsensical enough to stick a finger in a dragonfly's mouth, as I have done, will find that the teeth can deliver nothing worse than a good tweezer pinch. The dragonfly's mouth is built for catching and grinding up stuff like mosquitoes, at a rate of three hundred

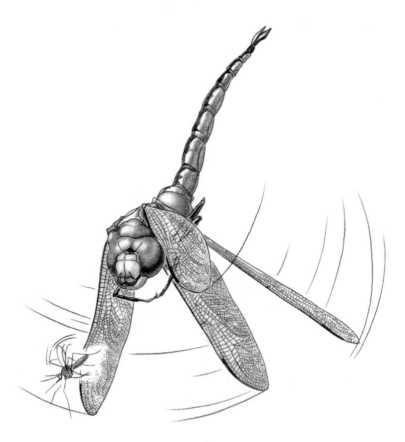

or more a day. "Mosquito hawk" is thus one of the few accurate common names.

This may be one reason the dragonfly is currently enjoying a social rehabilitation in this country. The Dragonfly Society of the Americas, formed in 1989, now has more than two hundred members. The most devoted of them dig ponds or plant old bathtubs in their yards as dragonfly breeding sites, and even New York City's Central Park now has a dragonfly preserve. (Japan, where dragonflies are a traditional symbol of good luck, now has thirty dragonfly sanctuaries, including one based on effluent from a sewage treatment plant.) At the Cape May Bird Observatory in New Jersey, the clouds of dragonflies passing through have at times been so thick that hawk-watchers could not pick out the birds. So the observatory now offers guided dragonfly workshops.

It has actually dawned on a few heretics that dragonflies are more colorful, more fiercely predatory, and ultimately more interesting than birds. You can get close enough to them to see what they're doing. With a net and good reflexes, you can also catch them and hold in your hand one of the most ancient and adept aerialists in the history of the planet—or at least you can swing and miss and know you have met your match in the animal kingdom.

One morning when I was just getting acquainted with the dragonflies around my New England home, I went to visit Ginger Carpenter of the Nature Conservancy in Rhode Island. Four of us, armed with nets, were soon patrolling knee-deep down the Queen River, where the Conservancy is conducting a population survey.

Carpenter, who is shorter than her net, button-nosed, with her blonde hair braided in back, has written a regional field guide to dragonflies, one of the few now available in this country. She's been chasing dragonflies for fifteen years. This means she has roughly 299 million years of catch-up work ahead. The dragonfly has been evolving its aerial agility since sometime before the pterodactyls. Fossils show that in its glory days, it had the wingspan of a crow. "There! There! *There! Gomphus!*" Carpenter yelled. She swung and missed. "Didn't get 'em."

The dragonflies made us all look inept. Lisa Barton, a Conservancy administrator I came to think of as "Lisa the Lunger," charged upriver with her net, plunging across the shallows and flogging the bushes as she sank in the mud. It was her first hunt but she managed to catch dragonflies. About all I could catch were jewelwing damselflies. They were everywhere, with their velvety black wings and metallic green bodies.

Damselflies are generally smaller than dragonflies. They are comparatively weak aviators, fluttering almost like butterflies where their dragonfly cousins cruise like helicopter gunships. When perched, dragonflies spread their four wings out to the sides; damselflies almost always fold their wings back over the abdomen. Dragonflies are also distinguished by huge wraparound eyes; in some species, the eyes actually meet in a seam at the back of the head. Damselfly eyes are more widely spaced, like a hammerhead shark's.

This stretch of the Queen River was narrow and gravel-bottomed, rushing quietly through the woods. I stood in the water, towering over a grassy patch where jewelwings danced lambently in the midday sun. The peaceable setting made their lives seem carefree and inconsequential. I caught one, then let it fly off my hand, and my entire being was suffused with the warm feeling I have come to recognize as Jolly Green Giant syndrome: Me big strong fellow, likeness of God; you little pissant. Yea, truly, I was green in the ways of the odonate.

Carpenter spotted a couple of jewelwings in a dogfight just above the surface. "You see those two males fighting over there? They both like the same female. She's sitting on the grass there. They're very aggressive. And there's quite a courtship dance between the male and the female." I watched the males bully one another, making dizzy loops and spreading their wings as if they could puff themselves up to grizzly-bear size and scare off the competition. The intensity of their lives began to dawn on me.

For a male damselfly, a suitable perch overlooking a few square feet of water is a territory. Defending it against relentless male competition is the best way to get a date. If he can hold a choice breeding site, the male may become what animal behaviorists call a "super stud." Females will dawdle past, sultry and coy. Scientists call this the "Lolita flight."

To court a female, the male faces rakishly away on his perch and trails his back wings down like a cape. He then lifts his hind end and shows off his colors. Finally, he flies up and hovers in the female's face, flapping his wings at four times normal flight speed. She presumably gets the message. By coincidence, jewel-wings (*Calopteryx maculata*) are the species in which a Brown University researcher named Jonathan Waage discovered the bizarre details of what happens next.

Before the male even gets started, he must, in effect, have sex with himself, transferring sperm from a storage reservoir at the tip of his abdomen to the copulatory complex at the base of the abdomen, near the thorax—a procedure otherwise unknown in insects, though it is common among spiders. During mating itself, the male latches on to the back of the female's head with the tow-truck apparatus at the end of his abdomen. He clutches a blade of grass with his legs and holds himself cantilevered out, supporting them both. She wraps her legs around his hind end and bends her body under his in a sort of loop, until her own hind end connects at roughly his midsection. Some species commonly fly in this "wheel" position, and it can be a romantic sight. Calico pennant dragonflies actually have a string of valentine hearts down the length of the abdomen, red in the male, honey gold in the female.

But Waage demonstrated that sexual aerobatics are the least complicated part of mating. During copulation, the male makes regular undulating movements of the abdomen. By dissecting couples killed *in flagrante delicto,* Waage discovered that these undulations are not geared primarily toward mutual gratification. Instead, they represent a methodical effort by the male to root out the seed of any males who have previously mated with the female. The male organ is equipped, like a Swiss Army knife, with an array of attachments minutely evolved for this task of copulatory dilation and curettage. Waage found that jewelwing males were 90 to 100 percent effective at removing the evidence of previous matings. Thus the male can be certain that the eggs the female deposits in his territory will be his own offspring. But males being males, doubts linger. Anyone who has spent time at a pond will have noticed one dragonfly flying with another in tow. In my ignorance, I'd always assumed they were enjoying some form of nonstop romance, an odonate waltz, round and round the pond. In fact, tandem flying is a form of desperate male clinging, called "contact guarding." The male hangs around lest another male swoop in and do unto his sperm as he has done unto his predecessors'. In some species, the male merely hovers possessively during egg-laying. In others, he remains attached and seems to swing his bride like a golf club as she pitches forth her eggs. The female presumably puts up with this because he protects her from the even more tiresome attentions of other would-be suitors. She gets twelve to fifteen minutes of undisturbed egg-laying, versus two minutes on her own.

The afternoon grew fat and hot, and the dragonflies zigzagged high overhead. "Look up there. It's a bunch of *Epitheca princeps,*" said Carpenter. "It's a big monster, two or three inches long. We call it 'the prince.' It's a beautiful thing. Look at them just cruising around, having a good time, picking off deer-flies. There's a squadron of them." She raised her net and begged. "*Come to mama*," she called.

They dipped and banked but kept their distance, and by mid-afternoon, having aggravated banged-up knees and bad backs with our log climbing and our Lisa-lunging, we abandoned the hunt, dreaming of beer and ibuprofen.

Despite their reputation as aerialists, dragonflies spend most of their lives underwater. In the larval stage, typically lasting one to three years, they are drab and, at a glance, unimpressive crea-tures, sprawling amid the underwater foliage or on the bottom. But they are also the scourge of all pond scum. A friend writes that as a boy he used to bring home a murderers' row from a local pond—water tigers, diving beetles, water bugs, back swim-mers, and of course dragonfly larvae. He'd start with twenty things in the aquarium, "and after a week there'd be only one thing, and it would be fat."

Dragonfly larvae have a secret weapon to ensure that they will be the one fat survivor: anal jet propulsion. They use the rectum to breathe water in and out across gills in the abdomen, and they are capable of ripping off a blast that will rocket them two or three inches through the water away from any predator. Damselflies are too refined for this happy talent; they have a more delicate three-finned tail assembly.

One day I watched a damselfly nymph scanning its under-water world with its big, wide-set yellow eyes. When a quarter-inch minnow approached, the eyes followed it, then the head turned, then the entire body swiveted around on its long legs. It yearned for the fish, the way a cat longs for a moth dancing just out of reach. But nothing came within striking distance.

But both odonates possess a killer lip. Normally, this hinged, movable lip stays folded underneath the head, where it looks like a beard, one-third the length of the body. But the larva can

unfold the lip and shoot it out in a hundredth of a second, seizing a fish or tadpole with two grasping hooks at the tip. Immature mosquitoes are among the favorite prey. In Southeast Asia, where mosquito-borne dengue fever is often fatal, public health workers recently discovered that dragonfly larvae are a good remedy. Two of them in a water storage container killed all larvae of the mosquito that causes the disease—up to eight hundred individuals in a week.

When they're done, if they do not eat one another, the dragonfly larvae do something pesticides often don't: They go away, and in great style. The larva climbs out of the water, splits down the back, and becomes a dead skin out of which the dragonfly emerges. While still soft and glossy, the dragonfly pumps itself up to stretch out its wings and abdomen, till it is often triple the size of the larval skin.

This is the most vulnerable stage in the dragonfly's life. At the Slaughter Ranch, I watched one gather its strength on some debris a few feet offshore. Then a black phoebe swooped down from a willow and gobbled it up for breakfast. But once it dries and hardens into its adult shape, a healthy dragonfly can usually outwing almost any predator.

When dragonfly enthusiasts meet over beer, the talk invariably turns to how some prized dragonfly has eluded them, and to new schemes for evening the odds. "I asked for a gun for my birthday," I heard an insect taxonomist remark one night. "I could load the shells with dirt and bring one down."

"I know hunters who have trouble hitting a duck," someone replied, "and you think you're going to hit a dragonfly?"

"How you gonna retrieve the specimen?" another doubter wondered.

"You need an odonate dog."

"If you do hit a dragonfly, you'll blow it to pieces."

"All I need are the genitalia," the taxonomist declared triumphantly, genitalia being the key to species identification. There was a pause during which everyone present considered the odds against training a dog to retrieve dragonfly genitals. Then the conversation shifted mercifully to other outlandish schemes.

Capturing dragonflies alive and unharmed used to be a common summertime game for boys in Japan. They tied small weights on each end of a thin silk thread, then threw the device in the air in front of a passing dragonfly. With luck, the dragonfly confused the weight for a mosquito, attacked, and got tangled in the silk thread. I chose a more new-fangled form of child's play, strapping Rambo-style bandoliers across my chest and wading out bearing a Super-Soaker XP 95 watergun loaded with soapy water. I'd been told this technique would temporarily foul the dragonfly's flying gear. Unfortunately, you must hit the dragonfly first. Every dragonfly I fired on saw it coming and ducked. I soon had a slick of soap bubbles spreading out grimly around me. Feeling like the captain of the *Exxon Valdez,* I quit and went home to read up on what makes dragonflies such elusive aviators.

Dragonflies are among the fastest insects on wings, having been clocked at thirty-five miles an hour. They can lift more than double their own weight, an achievement human aircraft cannot approach. They can take off backward, accelerate at warp speed, execute an unbanked turn in forward motion as if on a pivot, come to a dead stop in an instant, hover, and even somersault in the heat of combat. The U.S. Air Force has put them in wind tunnels to see how they do it, and despaired.

Dragonflies outpower all other winged insects; they devote one-third to one-half their body mass to flight muscles. But their flight system is one of the most antiquated on the planet. Each of the four wings is like a seesaw with an off-center fulcrum. One set of muscles attaches to the inside end. The other set angles out to attach to the wing on the outside of the fulcrum. The dragonfly rocks the wing up and down, achieving a modest thirty to fifty wingbeats a second.

By contrast, insects built on more modern lines, like the housefly, don't bother attaching the muscles to the wing at all. Instead, the first muscle set pulls down on a plate in the insect's back, causing the hard body surface to buckle inward and forcing the wings into their upstroke. The buckling simultaneously stretches the second muscle set, connected at the front and rear of the dorsal plate. These muscles then automatically contract, causing the body surface to pop back into its normal position and forcing the wings down. Houseflies thus manage two hundred wingbeats a second.

But dragonflies routinely catch and eat houseflies, not the other way around. So what gives? The dragonfly's wings, which look like brittle cellophane, are in fact intricately cross-braced, top-to-bottom and front-to-back, and it can bend them to produce almost any aerodynamic effect. It can beat all four wings in unison for quick bursts, or it can beat fore and hind wings out of phase for a slower, weaving flight pattern. Fine hairs on the upper wing surface, combined with the stubby antennae between the eyes, allow the dragonfly to read and control minute changes in airflow. On the leading edge of the wings, small thickened areas, called stigmas, may act as stabilizers.

All these adaptations make dragonflies masters of what aviation engineers call unstable aerodynamics. Whereas we strive to minimize turbulence over the wings of our aircraft, the dragonfly's wings deliberately generate and exploit turbulence. Wind tunnel tests of dragonflies showed a vortex over each wing, with the uplift of a miniature tornado.

Not only can the dragonfly outmaneuver almost anything else on wings, it can see better, too. Each of its bulbous wraparound eyes contains up to 30,000 lenses, and its field of vision extends to nearly 360 degrees. Dragonflies sometimes do things that call their visual acuity into doubt. For instance, species that have evolved along rivers often patrol highways instead, or they may lay their eggs not in a shimmery pool but on the polished hood of a car. But they can see other dragonflies and go to them from a distance of almost a hundred feet. A dragonfly perched in the grass can spot a gnat three feet away, zip out, grab it, then retreat to its perch to eat, all in just over a second, and do it 300 to 400 times a day. But let's cut to the bottom line: They could see me coming from a mile away.

Having failed thus far to catch a dragonfly, I headed out to the New Mexico–Arizona border, where the Dragonfly Society of the Americas was having its annual meeting. The spiritual leader of the group was Nick Donnelly, a sixty-three-year-old geology professor who looks like a parish priest. His white hair is scuffed up carelessly over a pink scalp. A paunch stretches his shirt front. When he speaks, he casts his small, downturned blue eyes heavenward, or closes them and purses his lips, as if almost tasting some ineffable dragonfly truth.

Donnelly started out as a birder, but lost his faith when he perceived that arguments over bird sightings tended to be won by loud voices, not hard evidence. Moreover, most of the species that birders in this country argue about were already known and studied more than a century ago. "Dragonflies," Donnelly said, "are an odd combination of really beautiful things, a number you can master—about five hundred species in North America—and the chance to discover something new. In North America, we're still finding on average one new dragonfly species a year. You just can't do that in the bird world or the butterfly world."

His wife, Ailsa, has followed him around the world in pursuit of dragonflies. "He says I should be down on my knees thanking God that he studies dragonflies," she interjected. "Because they live in nice clean streams. He could have gotten interested in the intestinal parasites of horses, in which case we'd spend every weekend looking at horse flop." She has wifely ways of evening the score. When "the big men go off in their big boots with their big nets," as she puts it, Ailsa finds a shady spot to read a novel, and en route often catches precisely the species for which her husband is out thrashing in the wilderness.

A dragonfly convention consists of forty to fifty fellow thrashers armed with nets who meet in a motel parking lot early on a Saturday morning, then scatter across the landscape. They reassemble at night and speak a language known only among themselves. When someone asks how big a species is, the speaker does not reply in inches or even centimeters. "About as big as *Erythrodiplax connata,*" he says, and everyone nods. Another speaker describes how he nearly died of hypothermia climbing a mountain in Papua New Guinea, and brought one prize back down—"a weird montane *Ischnura,* densely hairy and kind of fat."

Almost no one gets paid to study dragonflies, which do not damage crops, cause disease, or otherwise call themselves to the attention of agencies that fund research. So what we know about dragonflies is often the work of amateurs risking hypothermia in New Guinea, or getting their brains baked in the Arizona sun, usually on time stolen from their jobs, their families, or both. The members of the society are avowedly dragonfly *collectors,* not merely dragonfly-watchers. They go out equipped with glassine envelopes into which they carefully tuck the dragons and damsels they catch. They spend their nights preserving specimens in acetone, straightening abdomens and untangling thorny legs for a museum-standard presentation. Often the only way to know for sure what species you are looking at is to collect it and examine it under a microscope. "In the world of dragonflies," said Nick Donnelly, "you have the physical evidence and you can run with it."

"We're all trying to do serious science," Steve Valley remarked one day when we were out flailing our nets over the surface of New Mexico's Gila River. "We're not out stamp collecting. We're not trying to get as many as we can catch. It's not like birding where you're out trying to build up a life list, though there's nothing wrong with that. I do a lot of behavior studies, flight season things, geographical studies for my state. I'll mark dragonflies and go back to see what territory they're defending the next day and get longevity statistics."

In part because of the increasing interest among birders, the society is currently assigning common names to every odonate species in North America. Like bird names, they are appropriately colorful—gilded river cruiser, beaverpond baskettail, sandhill bluet, cardinal meadowhawk. But people like Valley worry that the birders may bring their hands-off ethic with them. The note of defensiveness in his voice is a response to people who frown on collecting any form of wildlife.

"It's a vertebrate idea carried over to invertebrates, and people don't understand that there's a difference," said Dennis Paulson, a museum director who is both a birder and a dragonfly collector. A female dragonfly may lay several hundred eggs a day, whereas a bird lays three or four in a year. Moreover, the dragonfly's adult life span is typically measured in weeks. "Every dragonfly you catch is going to die very shortly anyway. If you're looking at populations, and not individuals, there's no way in hell we could collect every dragonfly in a population. We're not good enough, and we're not out there long enough, and there aren't enough of us in the country."

I took my first swings out at a riffle on the Gila River at a dragonfly named *Brechmorhoga mendax*. I'm not sure why scientists gave it that name, which means "liar," except that it never showed up where it said it was going to be at the moment I started my swing. The other collectors advised me to wait till *mendax* was already flying past, then swing from behind, where its vision is weakest. But you cannot really blindside a dragonfly. Moreover, *mendax* always seemed to disappear just at the moment it flew closest. I tried to come in overhead, flapping the net down on the surface, a technique called "pancaking." I caught a fish. It was not a keeper.

"Play a lot of tennis to get the hand-eye coordination down," someone advised.

"And get a big net."

"Squash is probably the best game for learning to catch drag-onflies," another guy drawled. But I didn't notice them catching *mendax* either, though one of their wives had.

I went home empty-handed and sought out the wisdom of an old master, Steve Valley, who said, "When you get the net in your hand, you just want to hone that skill. It's like hitting a home run. But I miss most of 'em that I swing at. They've been doing this for a couple hundred million years, and nobody beats 'em for maneuverability. Think of all the visual data they've got coming in, and the flight corrections, and the stopping on a dime and then just disappearing. I'm out there because I enjoy being outdoors, and I love watching dragonflies fly."

So I relaxed a little. Next day, I was alone on the Gila River downstream from an area where the tan-and-yellow cliffs stood in columns like massive tree trunks. I was shin-deep in a riffle when a dragonfly the color of a flying goldfish cruised through at eye level. I watched it come and go, always just out of reach. It was a lovely sight.

Then it came straight at me and I crouched in readiness. It was like a baseball pitch, Mother Nature on the mound and in the groove. The ball had incandescent red eyes, which seemed to glare and taunt from fifty feet out. It moved like a slider and a curve ball and a change-up all combined. There was every possibility that it would stop six inches beyond the maximum extension of my net and zoom straight back in the opposite direction. I held the sock of the net against the handle, to minimize drag. The dragonfly kept coming. I swung, letting the white netting free, and listened to it whistle through the air. At the end of the swing I flipped the handle with my wrist so the sock hung down over one side. Then the dragonfly was buzzing angrily within my net.

I'd like to say I let it go. But I had faced it and won, eyeball-to-eyeball in its own territory, and I felt that it was mine. The scientist, or maybe it was just the hunter-gatherer in me, asserted itself. I found that I could catch different dragonflies with different strokes, and by the end of the day I had filled a dozen glassine envelopes.

The preservation process was a disappointment. The acetone held the enameled colors of the thorax and the abdomen reasonably well. But it could not preserve the magnificent eyes. They went from opalescent pools, in my favorite specimen, to the flat, opaque color of rust.

I expect I will catch and keep other dragonflies in the future. There are a couple of river estuaries near my house that have never been surveyed, and an entomologist at the state university wants to know what's there, with sample specimens. But I've come to love merely watching dragonflies fly. I love catching them to hold in my hand for a few moments. The eyes are misty and deep, like a fortune-teller's ball. Colors flash across them. Black patches like pupils seem to stare back, as if considering who I am and where I stand in the history of the planet. Looking into those eyes is like looking back in time. I remember that dragonflies were here before some of our ancient mountains. They witnessed the coming and going of the dinosaurs, the arrival of those second-rate aerialists, the birds, the evolution, just yesterday, of the human race. When I let a dragonfly go now, it no longer makes me feel like a giant.

Spider Love

Somewhere up the Amazon, a big pink river dolphin breaches the surface with a pneumatic venting of its blowhole, then glides off humpy and slow through the placid brown water. A flock of parrots brawls homeward overhead. The setting sun lights up the sheaves of tall grass on the riverbanks, into which we have nosed our boat for the night, and squadrons of mosquitoes wing down to join us for a drink.

Our expedition in search of tarantulas is traveling the upper Amazon on a long, thin, bathtub toy of a riverboat, painted ocher and green. An open-sided white cabin runs the length of the boat, and the ceiling is hung with clear plastic bags holding live snakes, lizards, scorpions, and, above all, large, hairy tarantulas. We make our bunks on the benches below. This is Peru, in the sort of terrain where an early adventurer, P. H. "Exploration" Fawcett, thrilled his readers with reports of a monstrous black tarantula that "lowered itself down at night on the sleeper beneath, and its bite meant death." As it happens, Fawcett later vanished in the jungle. We somehow live to tell the tale. Even the spiders that escape from our collecting bags leave us alone.

Tarantulas are by and large timid creatures. Like almost all spiders, they are venomous, but they rarely bite people, and the medical literature does not contain a single reliable report of a human death from the venom. The fear of tarantulas is so wildly exaggerated that our guide, a highly regarded tarantula expert named Rick West, interrupts any discussion of his favorite subject with a small, slightly defensive assertion: "Tarantulas are kind of boring," he says.

We humans, on the other hand, are genuinely scary. Besides West, our gang of four includes a writer, myself, with a special interest, somewhere between the scientific and the sociopathic, in animals humans commonly deem loathsome; a herpetologist whom we have learned to locate in the deepest jungle at midnight by his habit of belching like a frog (the frog, he explains, with unabashed scientific precision, would be *Hyla boans*); and an entomologist who aspires through long study and personal zen to achieve the worldview of an insect (walking one night under the green vault formed by a tangle of feathery ten-foot-high ferns, he exults, "I feel like a flea on a bird's back").

By coincidence, all four of us are large. We are bearded. We are hairy. On appearance alone, we are capable of frightening strong men, sensible women, and small children. We can empathize, on many levels, with the spiders we have come here to study. When we visit a remote village seeking tarantulas, the effect, West observes, is about what it would be like if a Viking horde were to descend on a North American backyard asking to see the earwigs.

But there is method in our madness: The Amazonian rain forest is one of the richest habitats in the world for tarantula species, many of them unknown to science. When we play our flashlight beams up the tree trunks at night, we can spot them here and there, camouflaged as lichenous pink starbursts on the mottled bark, poised for some unsuspecting insect or amphibian to come in range. When we tread softly, we can find them sitting and waiting at the mouths of burrow holes all over the forest floor.

In places, they are numerous enough to be among the dominant predators shaping life in the forest. A recent article in a scientific journal asks the obscure ecological question: "Why do Snakes Sleep on the Vegetation in Central Amazonia?" The author lists the danger of being eaten by a ground-dwelling tarantula as one likely answer.

As a newcomer to the subject, my interest in tarantulas is twofold. First is a question I contemplate as I lie in my berth with the spiders just overhead: Since tarantulas pose no plausible threat to humans, why does the merest glimpse of these creatures rile up so many primordial terrors? How is it that the mosquito can give us malaria and yellow fever, both endemic to the area, and seem like a mere nuisance, while tarantulas give us nothing worse than the willies and yet get typecast, in the words of the 1957 film *The Incredible Shrinking Man,* as "every unknown terror in the world, every fear fused into one hideous night black horror"? Considering how well we have succeeded as a species, human beings can be extraordinarily wrongheaded about recognizing our real enemies in the natural world. It would be a question to sleep on, if anyone were getting much sleep.

Instead, we spend the night out in the rain forest pursuing a second question: What, in fact, is a tarantula, and how do these strange and splendid creatures live in the wild? There are about eight hundred tarantula species in the world and they inhabit every continent except Antarctica. In this hemisphere, they range from Argentina to Missouri, and from remote rain forests to the saguaro and joshua tree deserts of the American Southwest.

The name tarantula itself comes from a cult in Taranto, Italy, where the bite of a spider was blamed—or served as pretext—for Dionysian revels of frenzied dancing and singing. The cult gave us a good dance, the tarantella, rooted in bad biology: Taranto has a kind of black widow spider, whose bite can be highly toxic. But the cult arose around a much larger wolf spider, which looks dangerous, though its bite is harmless. In the popular mind, the term "tarantula" has since come to mean almost any big, mean, hairy spider.

For researchers, it now refers not to wolf spiders, but to an entirely separate spider family, the Theraphosidae, mostly big, always hairy, and often with grooves on the carapace arranged like spokes around a distinguishing dimple at the center. They can be formidable spiders, living ten years or more, and growing to an average legspan of about five inches. A Venezuelan species is the biggest spider in the world, with a dinner-plate legspan and a rump the size of a ripe plum. (West once ate one, wrapped in leaves and roasted, and reports that it tasted like prawn. He used its fangs for toothpicks.)

If size is part of the tarantula's scary image, it may also be a blessing. Tarantulas are too big to stomp underfoot, the common human response to lesser spiders. They are, in fact, big enough for ordinary humans like me to see them as real animals, not merely objects for unthinking dislike. Watching tarantulas in Peru, and later in the American desert, and at home, where I acquired a tarantula as a pet, I began to see that spiders can be lovely: The velveteen fur, the plush cat feet, the high, arched legs, all eight of them, moving in delicate coordination, the subtle pink and brown and black colorations, the sinuous, finger-like weaving of the spinnerets laying silk.

One day I watched two huge tarantulas mating, and it had all the ferocity and passion of a tango. Gingerly, their front legs touched, then she sidestepped away, and he followed. With his pedipalps, the leglike appendages at his front end, he beat a tattoo on the ground, a declaration of interest. He began to court and caress her, drumming his pedipalps on her carapace. Gradually, face-to-face, they twined their front limbs together like the fingers of two hands in velvet gloves.

They pushed one another up in the reared-back position of both love and war. The male hooked his front legs over her fangs, and with his second set of legs held her down and bent her backward. Then he reached under to transfer the sperm from his pedipalps to the epigastric furrow at her midsection. Afterward, the male released one of the female's fangs and positioned his legs for an indelicate exit. In moments of postcoital *tristesse,* a female will sometimes kill the male, a handy source of protein for her newly fertilized eggs. This time, the dance ended with the male scrambling safely out of reach.

No one looks more closely at tarantulas than Rick West, who keeps two thousand of them alive in the basement of his home, plus another three thousand preserved specimens. He has been studying them without pay for most of his forty-five years. He is a pure enthusiast, earning his living as an inspector with the Society for the Prevention of Cruelty to Animals in British Columbia. He holds no academic position or degree, but museums and government agencies routinely consult him. West is also the favorite expert for the odd breed who keep tarantulas as pets (Chantelle, an exotic dancer, phones regularly about how to care for her pet Chilean rose hair). Tarantulas are a subject on which thoughtful and informed opinion is exceedingly rare, and sooner or later all roads seem to lead to Rick West.

Except that at the moment no roads do. We are wandering near the Rio Yarapa, ankle-deep on a soupy trail through the rain forest. "Pretty wet," West remarks dismally. He is pale and freckled, with close-set blue eyes, a red beard, a mournful manner, and a cracked, comical worldview. If tarantulas are so boring, I wonder, why keep a cellar full of them?

"Because they're different," West says. "Just when I think I've seen everything, somebody comes along with something new, some new color. My wife says, 'If you've seen one tarantula, you've seen 'em all.'

"I say the same thing about serial killer books. She buys every serial killer book that comes out, and she's absolutely fascinated by the psychology of the serial killer. So what's so bad about me collecting another tarantula? She doesn't see it, and I don't see it. But it's okay." It occurs to me that Tolstoy was being narrow-minded when he said that happy families are all alike.

West rolls over a rotten log and pushes aside some undergrowth in search of tarantula burrows. "Here I come, kickin' and flickin'," he announces, fair warning to any spiders that hadn't already been flooded out.

On a high spot, he finds a likely hole with a litter of desiccated insect parts nearby and starts to dig. Tarantulas are solitary creatures, and a single featherweight spider has dug this foot-deep burrow using only its mouthparts. West, who weighs 240 pounds, hacks with his machete into the gluey red earth. After ten minutes, he looks up, breathless and glowering. "It's abandoned," he announces. The spider has moved on and dug another hole somewhere else. A toucan passes overhead, and the sight fills West with fond longings for civilization. "God," he sobs, "I miss my Froot Loops."

When West says tarantulas are boring, he means in part that they don't use any fancy tricks to get their food. A tarantula cannot leap twenty-five times its body length to seize its prey, as some other spiders do. Nor can tarantulas construct elaborate webs, or hurl a sticky droplet at the end of a silken thread to lasso a passing insect. Exploration Fawcett to the contrary, tarantulas never lower themselves from the ceiling on strands of silk. "They're sit-and-wait predators," says West. "They don't do much."

Their venomous fangs are located at the front of the cara-pace, at the ends of two furry, fingerlike mouthparts known as chelicerae. Most other spiders bite with a pinching movement, like the grip of human thumb and forefinger. But tarantulas bite straight down, with a movement like the tapping of our index and middle fingers, enabling them to take on larger prey. Some tree-dwelling species can eat young birds in the nest. A large ground-dwelling spider can sometimes kill and consume a small rattlesnake or a fer-de-lance, one of the deadliest South Ameri-can snakes. But crickets, beetles, and other insects are more typ-ical tarantula prey.

One rainy night West pointed out a tarantula just inside the mouth of its burrow in the undergrowth, a home which at that moment looked as cozy as Mole's House in *The Wind in the Willows*. A giant cockroach, three inches long, entered the tarantula's neatly cleared little forecourt, an area West calls the "arena," and the spider began almost in slow motion to turn. Tarantulas sometimes stretch out strands of silk like a doormat to amplify any disturbance and announce the arrival of an in-truder. Like other spiders, they also have extremely fine sensory hairs on their legs. These hairs, called trichobothria, are set in pits with nerve endings on all sides, to locate the source of even the slightest vibration. The tarantula stepped out from its bur-row, then lifted two front legs to touch the cockroach gingerly, almost affectionately. The roach hesitated.

"It's as if she tastes it, to determine what it is," West whis-pered.

Then, in a blur, the spider latched its feet onto the roach's far side, flipped it onto its back, and planted its fangs in the rela-tively soft membrane of the underside, near the head. It dragged its victim into the burrow. The roach twitched briefly, then went still. The spider's fangs continued to rise and fall, pumping in venom. A clear liquid began to seep from the roach's farthest joints. We could hear the spider's fangs and the serrated teeth on the underside of the chelicerae begin to click like lobster picks at the shell.

Spiders cannot eat solid food. Instead, they pump digestive fluids onto their prey. Then they lap up the liquids. Another night, we watched a tarantula gradually open a gaping red hole in the belly of a small bat. West nudged the spider, which was perched on the trunk of a palm tree, into a better viewing position. "They're very single-minded about food," West said, when the spider made no attempt to escape. "He won't let go of that bat: 'This is *MY BAT.*' " The spider was supporting its own weight and the bat's on a sheer vertical patch of glossy green bark. We got close enough to study its plush footpads with a magnifying loupe. Tarantulas have retractable catlike claws, but they can also move nimbly on vertical surfaces because each hair on their feet branches out into hundreds of tiny bristles. "If you were to look at the bark of this tree under a scanning electron microscope," said West, "it would appear like giant cracks and craters and pits. So it would be easy for those hairs to find a place to hold on." Occasionally, when an insect annoyed it, the tarantula lifted a leg and shook it like a dog, as if oblivious to the normal laws of gravity.

With the bat clutched underneath, the spider began to rotate clockwise. Dewy strands of silk emerged from the spinnerets at its hind end, and wrapped around the bat's leathery, striated ears and over its eyes, which had started to ooze.

"As the tissues dissolve, the limbs will come apart," West explained, "and the silk holds it in a neat package." The bat carcass lifted and fell with the slow pumping of the fangs. "The spider keeps injecting digestive juice into the tissue to make it dissolve the meat. At the end of the night, the only thing left will be the wings, the bones, and some hair in a big pellet mixed with silk, and it will drop it to the ground."

The tarantula's abilities as a predator partly account for its gruesome reputation. In Southeast Asia, one tarantula species is known as the "earth tiger," for the speed with which it lunges from its burrow. In Central America, tarantulas are called "horse spiders" because of the mistaken idea that their bite can cause a horse's hoof to fall off.

From such exaggerations, it's a small step to a belief in the tarantula's almost supernatural powers. In Cameroon, for instance, traditional seers use a tarantula to foretell the future. The seer sets a container over the spider's burrow, and places an assortment of elaborately patterned leaf cards around the entrance. The spider rearranges the cards to yield such advice as "The young man's head is filled with evil" or "You possess a good hoe." But the wildest mythology of the tarantula as archfiend is a product of Hollywood science fiction. In the 1955 film *Tarantula*, for instance, a desert town faces a hundred-foot-tall, cattle-eating, house-crushing tarantula. Clint Eastwood plays an Air Force pilot who saves the town from hairy doom by dropping napalm on a spider "more terrifying than any horror known to man."

The sorry truth is that tarantulas are mere animals, vulnerable, like other predators, to the natural order. One afternoon, West called us over to see one of the spiders he had collected. She was lying on her back as if dead. Then, with an eerie, trancelike motion, her body began to swell and contract. A tiny split appeared on her flanks, where her eight legs were socketed into the body along the edge of the carapace.

All spiders have an external skeleton, which they must shed as often as four times a year when they are growing. To tear apart her old exterior, the spider forces blood out of her bulbous abdomen and, like Popeye flexing his muscles, pumps it over and over into her extremities and into the area under the carapace.

West pointed to the silken mat she had prepared for her labors. "That's to prevent ants and centipedes from attacking her while she's in such a vulnerable position. Normally she would be underground, and the mouth of the burrow would be heavily silked to keep out predators." In this hemisphere, many tarantulas nettle their silk with stinging hairs as a form of self-defense. A single spider may have a million barbed hairs on its abdomen, and under other circumstances it can kick these hairs up in a cloud to drive off predators.

By now, the tarantula was almost finished with her molt. Her top and bottom had spread apart like a biscuit rising in the oven. She began to draw out her new limbs, shrugging off the old skin. "See the bend in that new tibia, how soft it is?" said West. "They're just like rubber. There. Everything is loose. She's got her legs out." The spider and her shed skin, on which she now lay, looked like complete duplicates, except that where the old exterior was dog-eared, she now wore a glossy gray velvet coat, on which guard hairs stood out in brilliant red. She continued to swell and contract, to get some growth while her new skin was still soft.

"It's one of the most energy-consuming things they do," said West, who sat to one side like an intern in a maternity ward, attentive and a little tired, conscious that the patient was doing the real work. "Sometimes if she doesn't get enough nutrients, she won't have the energy to untangle herself and she'll die, half in, half out, trapped in her own skin."

A tarantula can be an irresistibly rich source of protein. Among their known predators are certain species of storks, owls, lizards, and snakes. But the most impressive is a huge

Pepsis wasp called the tarantula hawk. Not long after we returned from Peru, I headed out to Arizona to see what tarantulas are up against.

As in the science fiction film, tarantulas roamed the desert. But they were small and gentle enough that I could cradle them in my palm without risk. Most were males, out wandering at dusk and dawn in a desperate search for a willing mate. Oblivious to the mundane business of eating, they had shriveled away to little more than legs and sex drive. The wasps generally ignored them, preferring to hunt down the more robust female tarantulas in their burrows.

The aptly named tarantula hawk is typically about two inches long, with veiny, rust-colored wings and a metallic blue-black body. Its lanky, articulated legs have twin barbs halfway down, and they end in hooked claws, for grappling with the tarantula. The stinger at the end of the female's abdomen can be up to a third of an inch long, and a government entomologist who is a connoisseur of insect stings told me that it is as impressive as it looks. "The *Pepsis* wasp sting," he said, as if savoring the memory, "is kind of . . . profound. It's not like these things that make you swear and jump up and down and say bad things about somebody's mother. These things, when you get stung, you might as well look around so you don't hurt yourself, and then lie down and scream. Why not? It takes your attention off the pain."

Patrolling among the saguaro and mesquite, the wasp finds a tarantula burrow and teases the guard silk at the entrance, possibly imitating a male tarantula's opening serenade. If that fails, the wasp will actually enter the burrow to draw out its prey. The first time I saw this happen, the tarantula erupted out of her burrow and reared back in the classic posture of attack: front legs up for the strike, pedipalps elbowed back, fangs flicked out, a blaze of orange hair visible just underneath like a gaping maw.

The spider reminded me of some silent movie sorcerer, body bent back in a malevolent curve, arms arched high overhead as if to sling forth bolts of evil magic.

The wasp's bold strategy is to slip directly under the venomous fangs and plant its stinger in the soft tissue of the tarantula's hind end. The effect on the tarantula is immediate paralysis. The wasp then drags it off to bury as a macabre nursery for its offspring, laying a single glistening white egg on the victim before covering it. When the egg hatches, the single wasp larva will dine on the living tarantula, avoiding the vital organs at first so its immobilized food supply will remain fresh for as long as a month. It is the sort of thing that can make ordinary people actually sympathize with tarantulas. Once or twice a year at the Sonora Desert Museum outside Tucson, a visitor will drive off the wasp and rescue the paralyzed tarantula. Staffers accept the victim politely (never mind that the wasp has surely earned this spider) and nurse it back to full health.

The path from fear of tarantulas to sympathy and even affection may be a peculiar one, but I'd found in my own family that it was surprisingly profound. My children had dubbed our six-inch Chilean rose hair Queen Mary, and my wife, who is generally dismayed by the creatures with which I associate on the job, cooed over the tarantula from the start. Queen Mary was, in truth, the perfect pet—quiet, furry, at home in an ordinary ten-gallon terrarium. She ate only crickets, and never bit the mailman. These are traits that have begun to make tarantulas increasingly popular as pets. So much so that officials have had to regulate international trade in wild-caught Mexican red-knee tarantulas, a showy spider. Last year, a California man was convicted of smuggling six hundred of them into the United States with an estimated retail value of up to $130,000.

But Queen Mary's value was far subtler than that. She was pleasingly different from other pets. She had warm eyes, yes, but eight of them, in a tufted tubercle at the front of the carapace. I began to think, a little smugly, that the relationship of dog and master staring into each other's eyes was just another narcissistic mammal thing. But a tarantula's strange and placid life was a way of seeing into another world. Watching Queen Mary bound us to her as if with a silken knot.

I didn't realize how strong the bond had become until it became necessary for my family to pack the spider in a deli container and ship her to me overnight in the field, for purposes of research. When I phoned home the next night, the first words I heard, in an anxious tone I myself seldom elicit even when I am out in some godforsaken corner of the planet, were: "How's Mary?"

She was fine. The search for what makes other people so fearful of tarantulas took Queen Mary and me finally to a conference room at the Stanford University Medical Center in Palo Alto, California. A psychiatrist named C. Barr Taylor was treating a patient for arachnophobia, the fear of spiders. It can be a disabling phobia, with symptoms ranging from anxiety to an unwillingness to enter whole areas of one's own house. But Taylor had promised me that the patient would be handling a tarantula, every arachnophobe's nightmare, after just an hour or two of exposure therapy.

The therapy was remarkably simple, a gradual, step-by-step introduction to the reality of spiders. First, Taylor showed the patient a bulb-shaped drawing roughly corresponding to a spider's body, then a drawing with two legs added, and then all eight legs. "Ugh . . . no . . . it gives me a creepy feeling," said the patient, a translator named Tina. She looked away, then back again until her anxiety slowly eased.

It was up to her to decide when to move on, and at each step, Taylor performed the new task first to make her comfortable. He identified the spider's legs, head, and abdomen, and she repeated this anatomy lesson at each step—first on a black-and-white photocopy, then on a full-color photograph of a tarantula, using a chopstick as a pointer. The ritual naming of parts began to demystify spiders. Gradually, the two of them moved on to a dead household spider, then a live one, and ultimately to Queen Mary, in a clear plastic box.

"Oh, yuck," Tina said. She dropped her head forward to avoid looking. "Oh! That's real science fiction."

"Okay," said Taylor. "Let's look at it in the box and identify the body parts."

Gradually, the patient regained her senses. "It's like a minky kind of spider," she said. "Like it's wearing a little mink coat." It was about then that I noticed the wide eyes and trembling hands of a medical center press person at the far end of the conference table.

The leading theory about arachnophobia is that most humans have a biologic propensity to be alarmed, at first, by spiders. At some point in our evolutionary past, spiders presented a serious enough hazard that the starburst shape or eight-legged walk got coded into our genes as a threat. "The way it moves . . . ," said the press person uneasily, after Taylor had eased her into the patient's seat. "It's like there's this place in my brain . . . way back there." She'd already been through the first steps of exposure therapy, but her knees knocked and her heels trembled at the sight of a live tarantula in front of her. "It's like the shape of the spider is so ancient . . . I'm not making sense, am I?" she said, her voice going wiry.

What seems to happen with arachnophobics, said Taylor, is that they've never gotten past the alarming image. "They've never paid attention to spiders. They've never *looked* at them. And when they pay attention in conditions of safety, they get past the phobia. The brain is looking for the threatening stimulus of the particular spider, and as it goes through the checking process and doesn't find any, the fear goes away."

The press office woman, a forty-seven-year-old named M. A. Malone, turned out to be a classic tarantula phobic. When she was a child, her two older brothers had tormented her by sending their hands creeping spider-wise across the dinner table, and when the word "spider" was banned from the house, by talking about *Speidel* watches. As a treat, when she was about six, the boys took her to the movies. Then, in the dark, they slipped away from her side. The movie was *Tarantula,* and when the hundred-foot-tall monster appeared over the mountaintop, with Clint Eastwood nowhere in sight, the two brothers pounced on Malone from the row behind. Part of her psyche had never quite come back down from the balcony ceiling.

Tentatively, Queen Mary set one foot on her hand, and Malone took a short, sharp breath. The tarantula climbed up into her cupped hands, which sweated and shook. "I'm amazed," Malone said. "I'm fascinated. I'm waiting for it to bite me." She kept her eyes locked on the spider, which did not bite. Her trembling subsided. "Wow, just incredible," she said. "Absolutely incredible." Unobtrusively, Taylor got up from the table and moved away to a corner of the room. Malone continued to cradle the spider. "I wish my brothers were here," she said, to no one. After forty-five minutes, she was still holding Queen Mary, a little piece of eternity in the palm of her hands. The cure for arachnophobia, Taylor told me, is simply to shatter the alarming image just once.

But the scary image of tarantulas is everywhere, as if, in the heart of our unnatural cities, we still need the thrill of ancient fears. At the end of our time in Peru, Rick West and the rest of our gang of Vikings, plus assorted local curiosity-seekers, had piled into a pickup truck to visit a tarantula collector on the dirt road thirty miles outside Iquitos. At about twenty-eight miles, the truck died in a geyser of steam from the radiator. We hiked the rest of the way in, turning off onto a swampy trail. Clouds began to darken the sky, and distant thunder resonated beneath the sonar pinging of a frog, and the high *tu-who* of a rail. We met a woman short several teeth who was carrying a string of small fish and eating wild grapes. It was another twenty minutes, she said, to Nilo, the collector. We heard the waterfall roar of a rainstorm drawing near, and the winds began to blow up among the treetops.

Nilo was a friendly, enterprising man in his mid-twenties. He lived with his wife and three children in a thatched hut, next to a small farm plot cleared from the forest. He eked out their living by gathering tarantulas for a souvenir dealer back in the city, and he led us out into the downpour to demonstrate his technique. He wore a blue plastic poncho, and a baseball cap turned backward on his head. In one hand, he balanced a machete by the bare blade. In the other, he carried a stick with a sharpened ice pick at one end.

It took him about ten minutes to excavate the first tarantula burrow, hacking out the clay with angled slices on either side of the entrance. When the tarantula was finally cornered at the bottom, it made a desperate lunge, and Nilo gigged it through the carapace with the ice pick. He held the tarantula up for display, and it wriggled on the spear point, its milky blue blood leaking from the wound. He would sell it in the city for about a dollar, with fifty to a hundred other tarantulas killed in the same way and preserved in alcohol.

Our truck would not start again, and we spent that night in and around it with assorted live tarantulas, which seemed utterly innocuous by now, and a live coral snake in a clear plastic bag, which took some getting used to. All night people woke up from their bad dreams to ask, "*Donde es el naca naca?*" or "Where's the damned coral snake?" We tied the bag to the handle over the passenger door, where the entomologist found the direct eye contact disconcerting, then tucked it into the glove compartment, until someone concluded that the glove compartment probably had not been designed to be snake-tight. Then we heaved it with considerable relief onto the muddy road outside, until it occurred to us that we might now step on it in the course of our nocturnal wanderings.

Three days later, when we finally got to the airport, entrepreneurs were selling souvenirs to other tourists in line. A man named Lucho came up to offer me tarantulas at $10 apiece, pinned out in handsome glass boxes. I thought about buying one. It was the perfect image of the tarantula embedded in our genetic memory, the monster whose bite meant certain death. Something to thrill the neighbors back home. But the preservative had destroyed the natural colors, and the spiders, tarted up like Halloween knickknacks, all looked greasy and dog-eared. I had seen how lovely they could be in real life.

"Thank you, no," I said, and to a stranger in line, I added, "Tarantulas are kind of boring, don't you think?"

Leapers

A flea under a microscope looks, at first glance, a bit like Wimpy, the celebrated moocher in the Popeye cartoon who used to say, "I'll gladly pay you Tuesday for a hamburger today." Through the lens, the glossy amber freckle of flea comes into focus: It is a hog-backed, round-bellied, no-neck sort of insect, dim-looking, with combs and palpi drooping down over its mouth like a mustache.

The flea is of course also a moocher. Over one hundred million years, fleas have evolved, sometimes on bizarre lines, to find their victim, leap on board, slip deftly through feather and fur, and bite without being harrowed up by angry beak, finger, fang, or claw. The flea has made humans, along with a variety of other unfortunate birds and mammals, its hamburger.

Modern life makes this pleasant to forget. But for much of our history as an ostensibly civilized species, humans of all classes were routinely flea-ridden. The flea, in the words of one old poem, was "born to range the merry world / To rob at will the veins delectable of princes . . . To lie with ladies, and ah fairest joy, / On infants' necks to feed."

Moreover, by some quirk of human psychology, the flea became not just our parasite, but our companion, an amiable little vampire. The louse, its fellow vermin, was just a louse, a victim perhaps of human attitudes about creeping versus leaping. But the flea burrowed into our hearts. "Don't you like fleas?" an eccentric Englishwoman inquires, in one old entomology textbook. "I think they are the prettiest little merry things in the world. I never saw a dull flea in my life."

Today it is possible to go through life without seeing any flea, much less a dull one, largely because our houses are drier and more likely to be vacuumed clean. But even as fleas have become less familiar to the unbitten public, their way of life has opened up to the small, brave band of scientists who study them. For about a century now, we have known that the flea can transmit disease, notably bubonic plague, and kill us by the tens of millions. The merry flea thus ranks among the deadliest animals in the world. More than itching ever did, this knowledge has enabled scientists to learn the extraordinary details of the flea's way of life.

We now know that there are more than two thousand flea species, up from ninety-five known at the beginning of the century. We know how some of them perform the human equivalent of jumping over St. Paul's Cathedral—and that they will do it not once, but six hundred times an hour, if need be, for three days in a row. We know that this fleck of dust, this incarnation of all things trivial, possesses what has been called "the most elaborate genital organ in the animal kingdom," and a proportionately complex reproductive life.

We know enough about the flea's bewildering array of bristles, spines, and combs that an expert can look at a flea and tell if its host species is a hedgehog or a bat, is active by day or night, lives underground or in the treetops. Researchers can study a flea's genitalia, combs, and hair and determine where its ancestors came from fifty million years ago. Or as an artist to the great microscopist Antonie van Leeuwenhoek once remarked, on examining a magnified flea, "Dear God, what wonders there are in so small a creature."

In a lecture hall at the University of Bristol on the west coast of England, Charles Neville begins his talk on "the flying leap of the flea" with the practiced showmanship of a circus ringmaster. The entertainment to come, he promises, will touch on rubber, muscle physiology, stiff cuticle mechanics, high-speed cinematography, evolution, and behavior. Neville is a mid-fiftyish, Falstaffian academic with a gray goatee and a baggy v-neck sweater over his lingering paunch. He speaks with a dry, self-mocking humor that has been almost mandatory among flea researchers at least since Aristophanes ridiculed Socrates as a dreamer who spent his time measuring the jumps of fleas.

For a flea to jump far enough to get on board a passing cat or dog, Neville says, "you need to develop terrific acceleration, especially if you've got short little hairy legs." A flea lifts off at 140 times the force of gravity, whereas fighter pilots "are not too happy" if they experience much more than six G's.

The champion leaper among fleas can soar to an apogee of about ten inches, cartwheeling toward its host with its legs extended like grappling hooks. The flea at this moment becomes a sort of acrobatic assault trooper, flinging itself at the hairy, windswept flanks of its target. More often than not, it lands back on the ground, facing the direction from which it started less than a half-second earlier. Then it does the same thing over again.

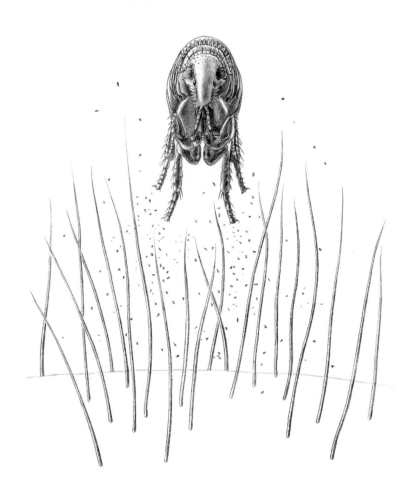

But even the fastest muscles known to science don't twitch fast enough to account for the flea's leap. So in the late 1960s, Neville became part of a team organized by British flea expert Miriam Rothschild to figure out how fleas do it. On the over-head screen, Neville projects a microtome slice of flea, with a spot stained bright red at its midsection. "The colored-in area is the resilin," Neville explains. He admires it a moment. "I love coloring in. It's the only thing I'm good at."

Resilin, an elastic protein, is the most efficient rubberlike substance known. A resilin ball dropped from 100 feet would bounce back to 97 feet (versus 85 feet for ordinary rubber). Resilin is common in insects, particularly as part of their flight systems. The thinking, says Neville, is that fleas evolved from winged insects, such as scorpion flies. But as they adapted to feeding on blood, fleas figured out that flying around looking for a meal was a waste of energy; natural selection instead favored fleas that waited patiently in the host's nest. Thus 74 percent of flea species are parasites of rodents, and another 6 percent, of birds. Here is how patient they have become: The Antarctic flea has evolved to wait nine months, surviving under several feet of snow and ice, for its host, a petrel, to come back to the nest. Fleas gradually lost their wings, which got in the way as they foraged through nests, feathers, and fur. But the resilin from the old flight system survived and became the driving force of the flea's flying leap from nest to host and back again.

A flea has a resilin ball sandwiched between plates of ordinary cuticle on each side of its thorax, above its back legs. "The flea crouches like an athlete in the blocks," Neville says, crouching for emphasis, "and the muscle contraction top-to-bottom squashes the two balls of resilin on right and left." The flea is thus cocked and ready to fire, with all the energy compressed in its resilin.

The trigger that sets it off is a single, central catch in its underbelly, which holds the thorax in the squashed-down position. "All you have to do is release the catch, and the energy is released from the legs," Neville says. He leaps, achieving an apogee of about one and a half inches. Actually, he notes, without further demonstration, the flea accomplishes its vastly more spectacular leap from its knees. Moreover, it can repeat its leap hundreds or even thousands of times an hour because it possesses what he calls the fastest single-twitch release mechanism of any organism.

But evolution is a wondrous and sometimes consoling phenomenon, and many flea species are no more adept at leaping than Neville himself. Fleas tend to be extraordinarily specialized. The flea that parasitizes shrews, for instance, would turn up its nose, if it had one, at a mole; likewise a swallow flea, even when starved, will not feed on human blood. Each flea species is also adapted to its host in the amount of resilin it possesses: A rat flea might overshoot its target if it jumped as high as a cat flea. Many bird fleas don't jump at all. Evolution has taken away their resilin altogether, says Neville. "Because if they jumped for the host and missed—*Bing! Wheeee!*—they would go all the way down to the floor."

Bat fleas also possess no resilin. But their young typically develop amid the guano on the floor of a cave. So they've evolved elongated legs to climb the walls back up to where the bats roost. One bat flea in Southeast Asia actually hitches a ride to the top of the roost on the back of an earwig; from a single earwig, researchers have recovered forty-one fleas, jostling and lurching upward like bus riders heading home to dinner—not hamburger, in this case, but a hairless bat parasitized by both the earwig and its passengers.

Despite all this specialization on other unsavory hosts, we humans don't have a flea species to call our own. Primates generally lack permanent nests where fleas can breed, and while humans are the exception, we are relative newcomers to domestic life, having moved into our first cave no more than two million to three million years ago. We are also adept at picking out itinerant fleas and cracking them between our fingers.

The twenty or so flea species that readily feed on people do so largely because of the company we keep. The so-called human flea, *Pulex irritans*, actually evolved to feed on pigs. The flea implicated in the spread of bubonic plague, *Xenopsylla cheopis*, feeds mainly on black rats. When humans complain about fleas now, they usually mean *Ctenocephalides felis,* the cat flea, which feeds promiscuously on cats, dogs, civets, hedgehogs, humans, and, in one case, even lizards. The cat flea is flourishing because it is more tolerant of our cleaner, drier modern housing. Meanwhile, the human flea has suffered a huge decline—the end of an arguably beautiful friendship.

To any cat or dog owner who has ever come back from vacation only to have the house spring to life with a *BOING,* it may seem loopy to suggest that humans now live in some sort of flea-free paradise. In the United States alone, fleas remain the maddeningly elusive object of a $1 billion-a-year business in flea collars, powders, pills, dips, shampoos, pesticides, and other products. But for the first time in our history, having fleas is a choice, an intermittent by-product of keeping pets, rather than an inescapable fact of human life.

The marks of our old forced intimacy remain embedded in the language. We go to flea markets, and our football teams execute flea flicker passes. We talk about flea-bitten ideas and fleabag hotels. But we seldom think about real fleas, and the itch of their incessant biting no longer pricks at our flesh, much less our sanity.

Flea-picking etiquette was never part of the schoolboy curriculum of George Bush or Bill Clinton, as it was for George Washington. (He had to copy out rules for the education of a French princess, advising good little children not to claw at their fleas in public.) "Flea brown" coats were never a fashion for Nancy Reagan. Queen Elizabeth II of England has surely never had occasion, as Queen Christina of Sweden did in the seventeenth century, to devise a miniature cannon for killing fleas.

So the modern mind boggles at the attempt to understand how intimate we once were with the flea. It is a bit like Zsa Zsa Gabor suddenly coming face-to-face with an odious stranger who turns out to have been husband number two (or was it four?). Who would imagine now, for instance, that women once wore flea traps in their undergarments, in lieu of more sporting forms of flea hunting? An eighteenth-century illustration shows a small ornamented cylinder hung from the neck, with perforations in the sides and a sticky pole in the middle to trap little invaders. Small furs also served as flea traps, and some writers have theorized that the lapdog became popular not so much to amuse the fond owner, as to divert the owner's fleas. It was the canine equivalent of a flea remedy employed by ancient Egyptians: Smear a slave with asses' milk and stand him in the corner of the room until well bitten.

The flea was such a standard part of our lives that it could serve even as our romantic surrogate. For all our well-known modern depravities, we may be too innocent now to imagine this. But men once wooed women with poetry about their fleas. A seduction poem by John Donne is the most famous: "Marke but this flea, and marke in this / How little which thou deny'st me is, / It sucked me first, and now sucks thee. / And in this flea our two bloods mingled be."

Lesser poets imagined themselves as a flea dallying across the soft landscape of a lady friend's person: "On a snow white breast I tarried a while / Then came to the place where pleasures smile, / For rashness now my life is ceded / Since here much bigger pricks are needed." Indeed, a whole literature of entomological erotica developed around the flea. The nineteenth-century *Autobiography of a Flea,* for instance, begins with the narrator "engaged upon professional business connected with the plump white leg of a young lady" and proceeds predictably downhill, and up, from there. Even the staid Victorian novelist W. M. Thackeray celebrated an unseemly delight in fleas in a limerick about a "perverted old man of Tabreez." According to Brendan Lehane, who gives an account of flea pornography in his book, *The Compleat Flea,* the flea served amorous humans as a convenient pretext, "an initiator of itch and fidget and ticklish desire." Also it was easy to rhyme.

Happily, science now suggests there was an element of truth in the flea's ribald reputation. Its reproductive life is complex. Fleas mate in roughly the spoon position, but with the female in back. The male slips under her abdomen and grasps her with his upraised antennae, then curls up his hind end to bring their genital openings into contact. His sexual parts, which occupy about a third of his body length even when coiled up within his abdomen, are equipped with spines, lobes, and, in some species, tickling devices resembling feather dusters or willow catkins.

Within the female's body, the male's sex organ must follow a meandering route, with the possibility for several wrong turns, to the point where the sperm will be deposited. Penetration alone can take well over ten minutes, and the male, with both their fragile parts at risk, employs a repertoire of tapping and stroking, as well as outright bondage, to get the female to stay put. The sweeping of his feather duster may also enable him to read her chemical signals.

A single act of copulation can last three to nine hours. It is an act worthy of an epithalamium sung by a chorus of virgins, but readers will have to content themselves with humming the tune to "Let's Do It," in which the noted invertebrate zoologist Cole Porter observed that, along with Cape Cod clams, katydids, and sentimental centipedes, "even educated fleas do it."

Sex among fleas is, however, even more highly educated than all of this would suggest. As part of their specialized adaptation, many flea species have become finely attuned to the body chemistry of their host. Some bat fleas, for example, seem to detect a seasonal change in the female bat's hormones, and climb on board just before she departs from her winter roost.

Most fleas mate only after they have first fed on blood from a suitable host, and rabbit fleas take this dependency to a remarkable extreme. They can't breed unless their host gets pregnant first. So the fleas become eager participants in rabbit romance. The moment boy and girl rabbit lay eyes on each other, their ear temperature shoots up, inadvertently signaling the resident fleas that the party is about to begin.

Fleas assemble on the pregnant doe's ears. There, ten days before she gives birth, they pick up a burst of hormones from her blood. These hormones trigger the fleas to reach sexual maturity just as the rabbit is about to give birth. A fresh burst of hormones causes the fleas to migrate to the rabbit's face and from there, as she nuzzles them, to her newborns, where a flea orgy is soon under way.

Over the next week, each female typically drops a total of about fifty pearly little eggs onto the floor of the nest. The adult fleas feed voraciously on the young rabbits and excrete a rain of half-digested blood. Their offspring need this food to develop through the larval and pupal stages. This is one reason fleas favor nest- or home-making hosts. Despite popular lore, fleas aren't everywhere, out breeding in the front lawn. They are far more likely to breed around the dog's blanket, or the cat's favorite rest spots, where their parents' droppings have accumulated. In the case of rabbit fleas, timing their reproductive cycle to the host's not only guarantees that juvenile fleas will be in the right place at the right time, with an ample food supply; it also provides a dispersal mechanism. The new generation of fleas grows up just in time to ride out for greener pastures on the backs of the young rabbits.

The researcher who described this extraordinary hormonal synchrony was Miriam Rothschild, and it put her in mind of the old lore that fleas attack women more often than men. Traditional sources attributed this preference to the more delicate skin and more sensitive nature of women. "Perhaps this is faulty reasoning," Rothschild remarked, "and the truth of the matter is that the human flea, *Pulex irritans,* also responds to the attraction of the ovarian hormones. This is food for reflection." Thus poetry and the romantic flea may have shown the way where science now follows.

What is an unfortunate flea-bitten female to do? At least according to lore, no animal is more clever than the fox at getting rid of fleas. When sorely bitten, she is said to back slowly into a stream, while clutching a patch of moss in her teeth. Her fleas scramble toward her snout to escape the rising flood. Finally, even the fox's nose descends, and the desperate fleas leap toward the safety of the moss, floating before them like a lifeboat. The fox then releases the moss, utters a satisfied "Ah," and swims back to land. It is a lovely myth. Unfortunately, it neglects to mention that when the fox returns to her den, a new generation of fleas will leap with a gratified "Ah" to reinfest her.

Real animals must work to get rid of their fleas, often gnashing desperately at their own flesh. Bad as it looks, this kind of grooming can be moderately effective at reducing the flea population. In one study, a mouse pruned its entourage down from fifty fleas to fourteen in a week. But when you scan the literature and see that researchers have collected 440 fleas from a single cat, 932 from a hedgehog, 7,000 from one miserable sheep, it becomes clear that fleas are also immoderately effective at holding on.

In Robert Traub's modest brick home in Bethesda, Maryland, the centerpiece of the dining room window is a stained glass image of a flea. Of necessity, it lacks most of the minute features of flea anatomy that enable the flea to hold on to its host, details which have fascinated Traub for much of his career in the military and later at the University of Maryland Medical School. "I worked on the epidemiology of insect-borne diseases and on control of disease vectors," he remarks. "Don't make it look like the Army paid people to investigate flea peckers." But for more than fifty years, even when his job took him in other directions, Traub has been obsessed with fleas—and particularly with their combs, bristles, and spines, their helmets and false helmets, their marginal tiaras and crowns of thorns.

Traub is a crusty figure in his late seventies, lean and bent, with a knife and a pen holder on his belt as if he were still out doing fieldwork in New Guinea. He wears big, black-rimmed bifocals, and a magnifying glass hangs from his neck. He shuffles down the narrow alleys of his basement, which is stacked to the joists with fleas. It is the second-best flea collection in the world, he says, after the one at the Natural History Museum in London that was built up over the past century by Miriam Rothschild and her father, Charles. Traub pulls open a drawer of a specimen cabinet and mutters, "Oh, this is an interesting one," then shuffles back to his bench with a flea neatly preserved in yellow Canada balsam on a microscope slide. He puts it into a microscope and projects the magnified image on a video monitor.

The specimen is a New Guinea flea that parasitizes a giant rat the size of a dog. The flea has a narrow, keel-like head for slipping easily between two adjacent hairs, and a bizarre extrusion called a crown of thorns at the back. The crown is perfectly adapted, Traub remarks, for hooking on to the hairs almost unshakably until the flea, in its own sweet time, is ready to move forward.

He puts up another slide and indicates a row of spines lined up at the back of a flea's head like teeth in a comb. "Flea people always acted as if these combs were there to help them with taxonomy," Traub says. "They never thought about whether they were useful to the flea." Charles Rothschild was the first to suggest that the combs were a means of hanging on to the hairy flanks of the host. "He presented the analogy that it was like throwing a rake into a tree. The teeth hang on to the branches and the rake won't fall down."

Over the past thirty years, Traub has taken the analogy much further, pointing out that, as part of their extreme specialization, each flea species seems to have adapted its rakes to the branches of its particular host. On his video monitor, Traub throws up a dizzying series of fleas to demonstrate: A flea with its comb teeth flared out at the tips to get a scissors-grip on the thick, coarse hairs of a hedgehog; a flea with narrow, evenly spaced comb teeth for the silky hairs of a mole; a flea with six or eight combs on its back, plus false combs and tibial combs, the better to hang on to a bat as it swoops and dives through the night sky, or migrates across half a continent. "As life becomes more hazardous," Traub remarks, "the fleas get more combs, more spines. We're at the point now where we can look at a flea we never saw before and tell you about the behavior of its host."

Traub also looks at fleas for evidence about the past behavior of the continents. The types of fleas on Madagascar, for instance, indicate to him that the island was once much farther north, probably off Somalia. The discovery that rabbit fleas in North America are attuned to their hosts in the same way as the rabbit fleas Miriam Rothschild studied in England suggests to Traub that this adaptation dates back fifty million years, to a time when Europe and North America were not so far apart. The similarity of the fleas on marsupials in Australia and South America led Traub to argue in 1968 that marsupials and their fleas island-hopped between the two continents via Antarctica. This apparently absurd idea, along with the whole theory of continental drift, is now widely accepted because of new evidence on plate tectonics. Traub's observation may take the prize in the category of seeing the world in a grain of sand. He mulls over a hundred million years of coevolution between fleas and their hosts as he browses through the drawers of his collection. He puts another slide in the microscope, caught up now in the wonder of fleas. But it is time for his visitor to leave.

On the other side of the ocean, in a dusty back room at London's Natural History Museum, a research assistant named Theresa Howard is addressing the only wonder most people think to accord the flea: wondering how to get rid of them. "This is a classic inquiry," she says, indicating a letter she has lately received together with a plastic bag of dog hair and a flea. " 'My dog has fleas. Here is the flea. What is it? How do I get rid of it?' I get these pulverized messes that used to be insects." But Howard is diligent. She wanders down the vast Rothschild collection of fleas, in their mahogany cabinets smelling of crushed naphthalene. Among other wonders, the collection includes a slide of a small plug of human flesh with a chigger flea at the center, neatly labeled "from inside the ball of the great toe of Capt. G. Hyde, S.S. Malvsen, 1896." The collection also includes a series of Chiclet-size boxes containing tourist souvenirs from a more flea-bitten era in Mexican history: *pulgas vestidas,* fleas dressed up as a miniature woodcutter, a very little drummer boy, and a bride and groom. Usually, the pulverized messes sent to Howard for identification turn out to be far more mundane; they are almost always cat fleas.

"I usually try to identify it and then I write a consoling note," she says. " 'You're not alone. It isn't because you keep a dirty house,' and I try to get them to see a vet. I tell them to spray—particularly where the animal sleeps. Boil the bedding, and if it has one of those classic old wicker baskets, scrub it out with a good solution of bicarbonate of soda in water. You have to be very tactful and say all animals have fleas. You just don't notice it until you have billions." She does not remind her correspondents that fleas were old hands at their business long before humans appeared on Earth. She does not elaborate on how good fleas are at what they do. She does not remind people how lucky we are, compared to our itchy, welt-riddled, flea-perturbed forefathers or, more often, foremothers. A degree of ignorance is indeed bliss. "The most important thing," says Howard, "is to be understanding."

Beetlemania

People who spend their lives studying bugs have an unfortunate reputation for being pretty odd ducks, and beetle collectors more so than most. On one field trip in Australia in the 1960s, for instance, two rival beetle enthusiasts set up their black lights for night collecting at opposite ends of the camp. One collector, fearful of being outdone, went to spy on the other. As he huddled in the shadows, a particularly rare beetle landed beside his adversary's light. The lurker stepped into the light, pulled out a gun from under his coat and declared, "*I'll* have that beetle."

So it ought to be said at the start that, in this context and by his own admission, beetle expert Terry Erwin isn't a wildly colorful character. There are times, it is true, when he will do colorful things: Asked to identify a bug that has turned up in the salad at his research station in southeastern Peru, he is likely to pluck it out and eat it. He has been known to persuade visitors that, in the Southern Hemisphere, the sun rises in the west. But by and large, Erwin is uneccentric. He has gotten where he has by solid and methodical study, concentrating on an unglamorous group of Coleoptera known as ground beetles. He has spent much of his working life not in the field, but at a dissecting microscope distinguishing among seemingly identical beetle species by such esoteric criteria as the hind angles of their crania.

The ground beetles have given him ample opportunity for making such distinctions. They are one of the most abundant and diverse groups of beetles, thriving in every continent except Antarctica and ranging in habitat from islands to mountaintops and from deserts and rain forests to glaciers. More than 40,000 species are known, and Erwin estimates that another 40,000 have yet to be identified. Outside his office in the National Museum of Natural History in Washington, D.C., where he is director of the biological diversity program, he keeps more than a million specimens in a long row of large, tomblike lockers labeled simply Caribidae, the scientific name for this family of beetles.

While God may have liked beetles, Erwin recognizes that most people don't. (In one novel, Joseph Conrad characterized an entomologist's collection of beetles as "horrid little monsters, looking malevolent in death and immobility.") When he describes a new species or suggests a change in taxonomic classification, he is acutely conscious that he is writing for an audience of perhaps four other experts in the world.

But Erwin's work has lately begun to take on ramifications for an audience well beyond this limited circle. The reason is

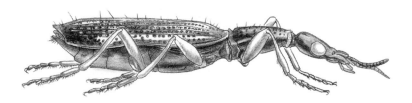

that he has made ground beetles an occasion for penetrating what he terms "the last biotic frontier": the canopy of the tropical rain forest. Up there, in the closely woven tangle of vines and tree branches, dwell snakes, land crabs, rodents and other mammals, insects, and birds that know little of terra firma. There are earthworms that never come to earth and ground beetles that would probably die on the ground. We know almost nothing about this exotic world. To put it in modern terms: Our nearest extraterrestrial neighbors are not in outer space, but ninety feet overhead. And they constitute an alien system.

Armed with ropes, a line-throwing gun, fogging machines, and collecting trays, Erwin has become one of the first researchers to penetrate this world and describe it systematically. Rather than merely discovering individual species and celebrating their wonders, he has cast his net to collect *all* insect species living in a given patch of rain forest. He has aimed to fulfill what botanist Peter Raven has called a need "to record coexistence and to be able to retrieve, for study, whole groups of organisms that occurred together."

Raven's use of the past tense suggests the pressing nature of the work: The alien world overhead will almost certainly perish in our lifetime. Raven has predicted that, if worldwide destruction of rain forests continues, one-fifth of all species on Earth— or above it—will disappear within ten to thirty years. One of the more remarkable results of Erwin's research has been to show just how little we know about what we are destroying. His studies have demonstrated that a single hectare of "unrich" tropical forest may contain 41,000 insect species (including about 12,000 kinds of beetle). The entire world, he says, contains not 1.5 million to 10 million species, as is generally estimated, but closer to 30 million. Many of them will disappear before we know what they are, which other species depended on them for their survival, or how they might have benefited mankind.

For Erwin, the study of ground beetles has become an occasion not merely for describing this vanishing world, but also possibly for saving it.

To reach Peru's Tambopata Reserve, where Erwin has done much of his fieldwork, it is necessary to fly to Lima, then over the ice-tipped Andes into a tropical pothole of a town called Puerto Maldonado. From there, a boat takes visitors four hours up the Tambopata River, which is a tributary of the Amazon. The landscape on either side is symptomatic: green, but unnatural—stripped of its forests and converted to short-term cattle-grazing and subsistence farming. The deforestation has advanced almost to the border of the twenty-five-square-kilometer reserve.

Biologically, Tambopata is one of the world's mother lodes. It harbors more recorded species of birds and butterflies (528 and 1,122, respectively) than any comparable place on Earth. Rare and endangered species, including the jaguar, bush dog, harpy eagle, black caiman, and giant river otter, thrive within its boundaries. One goes to sleep to the call of the great spectacled owl and wakes up to howler monkeys roaring like a distant wind. "Have you gone as far as the screaming pihas?" one visitor asks another, referring to an unmistakable point on the main trail where male birds of that species cry deafeningly for mates. Down on Laguna Chica at dawn, unseen birds make the whirping sound of a video arcade game. A kingfisher and its reflection skip belly-to-belly across the water. Then a gray-necked wood rail flashes down the length of the lagoon like a skater, shifting position in mid-flight as if for better display, or for the sheer pleasure of movement. It is possible, in a good half-hour, to see more species here than the Italian primitives dared put in their paintings of Eden.

To codify this richness and make human sense of it, Erwin set out to map one-hectare plots in each of the reserve's eleven types of forest. So far, seven plots are ready. Each is subdivided into twenty-five equal quads, with each quad containing fifteen to thirty-five substantial trees. Each tree is numbered and tagged. At the moment, Erwin and a crew of Earthwatch volunteers are thrashing through the undergrowth with measuring tapes. One volunteer calls out a tree's number and diameter at breast height, or "dbh." The others stretch measuring tapes from two corners of the quad to the tree, to define its position by triangulation. A remark Albert Einstein once made comes to mind: "The process of scientific discovery is, in effect, a continual flight from wonder."

The litany of numbers ("Point-1-9!" "Point 1-9?" "7-point-8-3!" "7-point-8-3?") is punctuated by the clicking of the keys of a laptop computer and the intermittent whirring of a disk drive. Computer specialist Margo Kabel has learned to practice her craft in the bush. Each morning, she and Erwin haul their electronic gear a distance of two miles along a muddy, boot-sucking morass of a trail. She sets up the computer on a tripod fashioned from an egg crate, with an umbrella propped gaily over the keyboard. She is careful not to get insect repellant on her fingertips, because it melts the surface of the keys.

Erwin uses the computer to construct a precise map of each quad and to update the map after periodic remeasuring. He hopes eventually to render each map in 3-D, so that the viewer can turn the image sideways and zoom in on a particular tree species or a distinct layer of the canopy. The idea is to correlate tree data with insect data, which Erwin and a succession of volunteers and assistants have been gathering methodically for years.

In the past, researchers interested in treetop life have by and large had to content themselves with whatever happened to fall at their feet. (These glimpses can be tantalizing. One day at

Tambopata, an Earthwatcher came back to the lodge with a dead wasp attached to a fallen leaf. Rootlike white tendrils emerged from its body. Erwin explained that the wasp had ingested a fungus spore, which grew in its innards until it burst through the wasp's thorax and feet, attaching itself and its victim to a leaf. It is the sort of thing the makers of the movie *Alien* had in mind.) To replace these glimpses with a panoramic perspective, Erwin deployed an insecticide fogging machine in the canopy.

Insecticides are of course notoriously unpopular with ecologists. Erwin chose a pyrethroid derived from the crysanthemum, then switched to a synthetic version that is said to have no effect on vertebrates and to break down within two hours into harmless substances. He also made his own trials showing that insects completely repopulated a fogged area within ten days.

Using ropes, he fogged at set points in each of his quads and repeated the foggings in successive seasons. Each fogging brought down a rain of insects, typically about seventeen thousand individuals per tree, to be swept from the funnel-shaped collecting trays into carefully labeled jars of preservative for subsequent study. Most of the insects were ants, which Erwin calls "the dominant driving force in the treetops." After one fogging of a single tree—"tree number 73"—in a single forest type, Erwin shipped off the ants to a colleague, E. O. Wilson at Harvard. Wilson identified twenty-six genera and forty-eight species of ants in the sample, an amount equal to the entire ant fauna of the British Isles. It was the largest collection of ants ever made in one place at one time. (Erwin has since topped that number. A tree he fogged at Manu National Park in Peru yielded seventy-two species of ants.)

While less numerous than the ants, the beetles were rich in variety: Scavengers like the darkling beetles, fungivores like the fungus weevils and the handsome fungus beetles, predators like the ladybird beetles and ground beetles of the genus *Agra* (the latter are Erwin's specialty and also the largest insect predators of the canopy; he describes them as jaguars of the insect world), herbivores like the scarab beetles, beetles that eat only rotting plant matter, and other beetles that eat nothing at all, at least in their adult stage (which is necessarily brief). Many of the species Erwin found were completely unknown. He named two sister species *Agra sasquatch* and *Agra yeti,* because they were big-footed. (Earlier, he had named a new species from Venezuela *Pericompsus bilbo,* after Bilbo Baggins in *The Hobbit,* because "it was short, fat, and had hairy feet.") A species that was difficult to describe because he had only a few specimens became *Agra vation.* Thousands of other unknown species from the fogging have not yet even been named or introduced to the scientific world, much less studied.

Beyond establishing the richness and variety of treetop life, Erwin intended to study its workings. Like most tropical biologists today, he wants to demonstrate that the rain forest is most valuable to the local community and the world as is, rather than as something to be slashed and burned for its lumber and arable soil. That means figuring out natural relationships that may be useful to humanity. It means finding plants and creatures that aren't merely strange and wonderful, but also potentially valuable in the production of food, pharmaceuticals, biological control agents, and other goods. Some of these unknown treasures may be beneficial simply as "germ plasm resources," according to Erwin, who writes: "Do we not need to know, in these days of genetic engineering, what the world holds in the way of genetic diversity and genetic variability?"

With this sort of thought in mind, Erwin looked down one day instead of up (his instincts as a ground beetle expert perhaps reasserting themselves) and said something like "Eureka!" He was standing in a swamp forest, which is by far the richest kind of forest in the Tambopata Reserve. According to his own surveys, land there sustains a hundred more trees and 30 percent more standing wood per hectare than in any other forest type. He bent down and picked up a handful of the black soil, which was visibly richer than the red laterite found in other forest types. Erwin's original thought was that it might be possible to skim this rich soil without damaging the forest, and use it in local gardens.

One reason tropical farmers must continually slash and burn is that ordinary rain forest soil quickly loses its nutrients; it becomes worthless for agriculture after a few years, forcing the family to move on. Erwin wondered if rich swamp soil might lend itself to sustainable agriculture over the long-term.

This was the beginning of what has become known as the Amazon gardening project. In truth, Erwin had no idea what made the swamp forest so fertile. (He is not an agronomist, after all. And he notes that human ignorance about tropical ecology is vast. On the one hand, 45 percent of all pharmaceuticals now in use are derived at least in part from plants. On the other hand, only about 1 percent of tropical plants have been screened for potentially useful properties. The figures for animals are even scantier.) The active ingredient in the swamp might be some soil microbe that wouldn't survive elsewhere, or it might be the swamp forest's self-contained drainage system, which tends to pile up rich organic debris instead of letting it wash down the Amazon drainage into the Atlantic Ocean.

Erwin decided to test the swamp soil in a raised garden patterned on Mayan techniques. Laborers from the nearby community of Infierno built the thing on a site within the reserve. Volunteers from Earthwatch and Britain's Operation Raleigh hauled buckets of swamp soil a distance of two kilometers to the garden; they filled half the garden with swamp soil and half with river soil, which is more accessible, as a comparison. An agricultural student from a Peruvian university oversaw the planting.

The resulting garden is both formidable and productive. It occupies a small clearing, a square roughly sixty feet on a side. The garden surface is three feet above ground level, held in place by a double stockade fence made of split pona palms. A drainage moat, in which fish and turtles will be raised for food and fertilizer, surrounds the garden. (The silt from the moat will also go back into the garden, duplicating the swamp forest's self-contained drainage.) When Erwin and the others go there to work, between bouts of insect-fogging and tree-surveying, they enter via a makeshift drawbridge. There is something a bit wishful about the place—it looks rather like a child's fort—and perhaps also about the idea of saving the world by raised gardening. But it also appears to work. If it can be made to work for local farmers as well—and that is the big question—Erwin believes it may at least be a way of saving Tambopata.

Terry Erwin did not have a naturalist's boyhood. His father was a builder-driver of sprint cars on the California racing circuit, and also raced airplanes, boats, and motorcycles. Erwin was at the track every night until he was ten years old: "I grew up in the seat of a race car." When he was fifteen, his grandparents gave him a Henry J, made by Kaiser-Frasier. He and his father stripped it down to a hot rod and spent the next five years drag racing around California. One looks in vain for early entomological leanings: Erwin became a Boy Scout, but did not win his nature merit badge; he belonged to the only branch of the scouts designated an auto club. He spent summers fishing fast Sierra Nevada mountain streams with his grandfather, and got his first limit—fifteen fish—at the age of five. But he used worms, not flies.

In college, at San Jose State, he started in engineering and lasted through second-semester calculus. Then he switched to life sciences, with the idea of becoming a dentist. His undoing was an elective in entomology: "I got excited, got off probation, and became a straight-A student." It was a clear case of role reversal. Erwin's uncle, an accountant, had two boys who became drag racers; Erwin became an entomologist. His father took to describing him as "my bug mechanic."

At San Jose State, two friends in entomology became interested in longhorn beetles, "which are very showy—they've been on stamps." Erwin got ground beetles, which haven't. He stuck with this specialty, going on to study with the leading carabid researchers at Harvard, in Canada, and finally in Sweden. When he joined the Smithsonian's National Museum of Natural History in 1971, he wrote up an ambitious proposal to study the ground beetles of California. But tropical research was opening up then, and seeing an opportunity, a superior crossed out California and scribbled in Panama instead. Erwin was on a plane headed south less than two months after returning from Sweden. It took him two years to figure out what had happened, and by then he was hooked.

If science is, as Einstein said, a continual retreat from wonder, it is also often a way of circling back to it again. For Erwin, the process began in a willingness to gather and consider huge quantities of information with little or no idea of where it might lead. He adopted a kind of blindness to results as a way of eventually seeing them clearly. "It's ninety-nine percent gathering data," he says. On the wall of his office in Washington, he displays a favorite quote on this point: "The expectations of theory color perception to such a degree that new notions seldom arise from facts collected under the influence of old pictures of the world." Some scientists take a hypothesis and see how well the world fits it; Erwin aimed to view the world neutrally, without hypothesis. Twenty years spent peering through a microscope at the hind angles of crania, or five years spent fogging the rain forest canopy might then add up to what Erwin terms "baseline data," the stuff on which observation, synthesis, theory could begin to take shape. This is where the wonder comes in.

One night at Tambopata, a forestry professor, Carlos Reynel of Peru's Universidad Agraria La Molina, delivers an informal lecture on his country's prospects. Out of a total population of twenty-one million people, he says, 50 percent are under the age of fifteen, and 75 percent are under thirty. There is little land to feed them; the coast is largely a desert, and the interior is mountains and rain forest. The rapidly expanding population is beginning to spill over the Andes into places like Puerto Maldonado, and up the Tambopata River, entering the rain forest as their own last frontier. Their immediate need is for land, not trees.

Reynel is in a sense irrelevant to this need. As one of Peru's seven plant taxonomists, he is studying the family of trees that includes figs and mulberries. But his research has turned up more than thirty forest species, formerly regarded as having no value, which in fact bear edible fruit. With hybridizing (much as early Peruvians hybridized the potato from a wild plant), some of these trees might feed people. It requires only time and research: During one weekend in Tambopata, he has already found twenty-one members of this family, including one not previously known in Peru. "If we Peruvians don't know our own things," Reynel says, "we cannot take care of them."

Erwin believes his own research will prove its value when that kind of hybridizing begins, and when Peru begins to manage its rain forests as its most valuable natural resource. At some point, for example, Reynel may need to know what insect pollinates one of his tree species. He will then be able to consult Erwin's list of insects found in such trees in each season. A lumber company, having taken a tree species from the rain forest and turned it into a plantation crop, may suddenly need to understand the ecology of wood-boring beetles. From Erwin's foggings, it will be able to learn which parasitic Hymenoptera (typically small wasps or ants) lived with these beetles in nature and kept them under thumb. (Such parasites are already widely used in this country as biological control agents for citrus and other crops.) A toxicologist may need to find natural antidotes to certain poisons. He will be able to consult research, now being undertaken by Erwin and a graduate student, into how tropical willow trees use toxins to protect their leaves from herbivores and how, in turn, certain beetles have learned to detoxify these leaves in their digestive tract so they can eat them with impunity. This same research may help agronomists understand—and perhaps manipulate—the ways plants adapt their defensive toxins or turn them on and off to control pests naturally.

"We have to support the idea," Erwin says, when Reynel has finished speaking, "that basic science can be turned into applied science, and that this applied science can support human need."

On one of his last days in Peru before returning to Washington, Erwin sets out by boat to visit the nearby village of Infierno. He is accompanied by Antonio Del Aguila, the student who is overseeing the Amazon gardening project. The aim of the trip is to carry the message—that science can support human need—directly to the people who surround the Tambopata Reserve. Erwin is bringing a token of friendship, a framed drawing of a new species from the Tambopata canopy. He has named it *Aspasiola eshige*. The species name, which means "rainbow," comes from the language of the Esse-Ejja Indians, who were the original inhabitants of the area.

But the trio is also bringing something more practical: detailed drawings of the Amazon garden, and a proposal to build a model garden within the community. On the trip downriver, Erwin talks about the prototype at Tambopata. Just six months old, it has already shown its worth by producing a crop during the rainy season, when most farmers leave their water-logged fields fallow. The secret seems to be the drainage system, rather than the swamp soil. He refers to it now as a "containerized" garden, and he believes that the prototype, which is less than a quarter-hectare in area, will feed four families.

There have of course been problems and surprises. Opportunistic fish have used a seasonal stream to climb from the river up into the moat; but they aren't necessarily fish of a desirable species. The pona palm stockade on one side of the garden has fallen off, necessitating the use of buttresses. But now snails and small rodents are using the buttresses as a route into the garden. Steve Maber, the engineer who designed the prototype, has tinkered with his plan to beat such problems. He's also come up with a one-family Amazon garden that does away almost entirely with buttresses and pona palm stockade. This new model should also be easier to build.

"A lot of people are dubious," Erwin concedes. "It's so new and strange. It may be that nobody will buy it. Or maybe it'll hang around in the literature till things get so bad people turn to it out of desperation. The only thing we can do is to show them how well it works, show it to the young people. They're very practical people. They're conservative, but when they see a good thing—like farmers all over the world—they'll do something similar."

At Infierno, the president of the community does not show up for the planned meeting. He and his subordinates are off in Puerto Maldonado for the day. No one is terribly surprised. Life here is perhaps more politically complicated than in Washington, and it turns out that Erwin has made his approach through the wrong middleman. Instead, he presents *Aspasiola eshige* to the principal and teacher of the community school.

Then Del Aguila explains the Amazon garden. It is a conversation punctuated by the words *clara* and *pero:* "It's a great idea, but . . ." The principal is skeptical that the disputatious town fathers can agree even on a good thing. Still, they strike a tentative deal for Erwin and his Earthwatchers to build the model garden on the school grounds in September.

On the trip back upriver, Erwin points out vermilion flycatchers and violaceous jays. Six red macaws with foot-long tails squawk across the river overhead. He and Del Aguila sip beers and discuss their day. Then, at the beginning of the Tambopata Reserve, he spots a boat tied up; its owner is at work in the reserve with a chainsaw, cutting lumber.

"I'd like to put his motor at the bottom of the river," Erwin says. But there is more resignation than racetrack machismo in his tone. The chainsaws are already gnawing away at three sections of the reserve, and there is no enforcement staff to stop them. Two days later, after Erwin has headed home, a particularly brazen woodcutter starts up his chainsaw within earshot of the lodge where the scientists take their meals. Armed with machetes, Maber and a few others confront him and seize the saw. Afterward they calculate the damage: he has taken down one sizable tree. But it has fallen into two of the quads that Erwin has painstakingly surveyed. Six numbered trees are gone, and twelve others are damaged. Back in Washington, Erwin enters the information into the computer and describes it flatly as the result of a tree fall.

It is a reminder—as if he needs one—that time is short, and the struggle very long.

Grunting for Wigglers

It's a moment charged with nature's drama, akin to David Attenborough creeping up on the mountain gorillas of Rwanda. I am thrashing through the piney flatwoods of the Florida Panhandle, into the secret realm of the great shaper of the earth. Listen! *Runt*-runt, *runt*-runt, *runt*-runt. It's a sound like a wild hog wallowing, somewhere out here in the Apalachicola National Forest. Over that way!

A few steps more and, through the burnt gallberry brush, I spy a man, grinning and sweating, slashed head to toe with soot. He is on his knees in front of a black stake, called a stob, hammered into the ground. He takes a shiny automobile leaf spring and strokes the flat surface down rhythmically across the top of the stob: *Runt*-runt, *runt*-runt, *runt*-runt. He pauses, listens, then varies the rhythm, catching the stob on a backstroke. The earth quakes around him. It looks like some primitive cargo cult ritual. In fact, the man is calling earthworms—"grunting" for worms, as they say hereabouts—and he is making a living at it.

As he works, Ruben Hill's glance darts around the floor of the forest, which the rangers have burned lately to knock down the undergrowth (incidentally improving visibility for grunters). All around, worms are poking their angry pink heads up through the blackened soil. They shoulder themselves out of their burrows, shimmying, dancing, twitching along on their ten-inch bellies, driven to frenzy by this infernal vibration. Then Hill, a forty-year-old who learned worm grunting at his daddy's knee, grabs a paint can and strides around snatching up his haul, about 160 worms in five minutes. They will do their final dance on a fishhook, having been gussied up for sale by bait shops under names like "Louisiana pinks," a cunning way to hide their true origin from interlopers, or simply "grunt worms." (Scientists called the species *Diplocardia mississippiensis.*) Hill eyeballs several wriggling canfuls. "I've made it to twenty-five hundred," he says. The early morning sun is still glinting sideways through the forest. "That's seventy-five dollars right there. Not a bad payday." He'll double it by noon.

We are talking serious worms here. Indeed, by now, you may well be worried that I am going to go on talking serious worms for the rest of this chapter, and could there be a duller animal on the entire planet? Possibly not. But the lowly worm is a subject of compelling fascination (or anyway, a square meal) for a surprising array of predators normally associated with swifter game, including grizzly bears. Worms were also a lifelong interest for no less a naturalist than Charles Darwin, who once wrote, "It may be doubted whether there are many other animals which have played so important a part in the history of the world. . . ."

So stop your whining, gentle reader. We are talking about a conundrum. Worms are a dismally humble creature on which we loftier earthlings nonetheless regularly pin some of our most grandiose hopes and emotions. Though they are common as dirt (or maybe because of that), worms glitter like pink gold in the eyes of rural entrepreneurs, who hawk them in hand-painted backroad signs ("Crawlers 'n' Plywood Lawn Ornaments!") all over North America. To environmentalists and government agencies, worms are the stuff of green pipe dreams: They eat garbage, turn it into precious compost, and grow fat enough in the process to be ground up for livestock feed. At least, that's the theory. Worms provide peculiar sustenance even to fly fishermen, who spend inordinate amounts of time stoking their vital sense of superiority with snide references to "garden hackle."

Before Darwin, it was all much simpler: Most writing about earthworms consisted of advice on how to get rid of them. Worms were the original "vermin," a word they contributed to the language (from the Latin *vermis*, or "worm"). Gardeners hated them because they made the footpaths unsightly. Farmers believed worms devoured their grain and spoiled their fruit.

People also disliked earthworms for less practical reasons. The definition in the *Oxford English Dictionary* is a compendium of traits the Western mind found unsavory: "A slender, creeping, naked, limbless animal, usually brown or reddish." "Worm" became a generic name for any contemptible life-form, from a tick to a tax man, including the serpent in the Garden of Eden. "O Eve," Adam lamented, in Milton's *Paradise Lost,* "in evil hour thou didst give ear To that false Worm."

The worm's subterranean ability to insinuate itself anywhere, unseen, made it especially suspect. An assortment of illnesses and mental derangements were said to be caused by worms, perhaps in part because of confusion with assorted wormy parasites. A worm was a worm even when it was a toothache, a malady said to be caused by a worm at the root of the tooth.

The worst thing about worms, of course, was the gnawing idea that the entire human body was doomed to become "worms' meat" (a phrase you can blame on Shakespeare). In the popular mind, earthworms would ultimately unravel each of us to the bone. Maggots and microorganisms actually deserve greater credit for this chore. But earthworm defamation lingers even now in schoolyard takeoffs on the old rhyme, "The worms they crept in, and the worms they crept out, And sported his eyes and his temples about."*

Among the earliest writers to question popular antipathy toward earthworms was the eighteenth-century British clergyman Gilbert White, in *The Natural History of Selborne.* Though worms might appear to be "a small and despicable link in the chain of Nature," he observed, they "seem to be the great promoters of vegetation . . . by boring, perforating, and loosening the soil . . ." and by piling up vast amounts of fine manure for the very crops they were thought to destroy. There was perhaps a hint of distaste in his remark that earthworms "are hermaphrodites, and much addicted to venery," or sexual license. (A Freudian would probably find traditional attitudes toward earthworms laced with sexual subtexts. Chaucer used the idiom "naked as a worm," and in folk medicine, powdered worms were a remedy for impotence.) But White also wrote, unequivocally, "The earth without worms would soon become cold, hard-bound . . . and sterile." It took more than a century for the world to catch up with this idea.

Charles Darwin's last book, *The Formation of Vegetable Mould,* published in 1881, was the first methodical study of earthworms. "We refer to it as the peak, the *culmination,* of his professional activities," a modern worm scholar enthuses. This may be a matter of perspective. But Darwin's worm book was surprisingly popular; he found himself besieged by

---

*Worm researcher Sam James gives this charming rendition: "The worms crawl in, the worms crawl out, / The worms play pinochle on your snout. / Your eyes fall in, your teeth fall out, / Your brain turns into old sauerkraut."

correspondence from backyard philosophizers of the worm. He also heard from critics who doubted—as they had doubted his work on evolution—that small causes could produce the vast effects Darwin attributed to worms.

Darwin came to recognize that worms are incredibly numerous, citing one estimate (modest, in retrospect) that 53,767 worms per acre foraged through an ordinary garden. He also demonstrated that worms transform the Earth by eating their way, over and over, through the entire upper soil layer. A worm can consume about 30 percent of its own weight daily in plant matter, minute animals, and dirt, much of which gets excreted on the surface in the familiar, coiling heaps known as worm casts. From the casts at one site, Darwin calculated that earthworms brought eighteen tons of soil to the surface per acre each year. That number may sound preposterously large, but it falls well within modern estimates. In the Nile delta, an extreme case, earthworms annually deposit *one thousand tons* of casts per acre—a weight equivalent to five hundred junked cars, but considerably more useful to farmers.

Hence the big question: Where does all this worm poop go? Darwin figured that, in his samples, it accumulated on the surface in a barely noticeable layer a fraction of an inch thick each year. But over three decades, he watched a flinty patch his children called the "stony field" become loamy enough to gallop across on horseback. Darwin was also the first to note the debt that archaeology owes to earthworms: The casts they heap up bury and preserve even the most massive objects, which the worms may simultaneously undermine with their burrowing. He attributed the sinking effect of the monoliths at Stonehenge, for instance, to worms.

Above all (or perhaps below), Darwin won popular acceptance for Gilbert White's idea that worms make the ground suitable for plants. They aerate and drain the soil with their burrows, drag down leaf fragments to grind up for food, and "mingle the whole intimately together, like a gardener who prepares fine soil for his choicest plants." Long before the advent of the plow or of the human race itself, Darwin wrote, "the land was in fact regularly ploughed" by earthworms.

Researchers now estimate that there are more than four thousand earthworm species plowing their way across the face of the planet. According to Sam James, a prominent worm taxonomist, the list grows by dozens of new species each year. They may vary widely in habit: Some live in rain forest treetops, for instance, and dine on decomposing material trapped in bromeliads at dizzying heights. Some may burrow eight feet underground. Some dine on other earthworms. A South African worm puts up such large casts that farmers must use bulldozers to level their fields, and an Australian species, the Giant Gippsland earthworm (*Megascolides australis*), can grow to more than forty inches in length. (James adds the important caveat that some people literally stretch the truth. Weight is a more reliable measure, and at the moment the purported champion is a Brazilian worm weighing about a pound.)

For most of us, though, worms are still worms. All species conform to the same efficient if uninspiring body plan: A tube within a tube. Aristotle described them aptly as "the intestines of the earth," all guts, no glory.

The best-known worms, the nightcrawlers (*Lumbricus terrestris*), are not native to this continent. They arrived here in the soil of the potted plants that early European settlers brought with them, and quickly came to dominate the vast northern zone left worm-free by ice-age glaciers. A crawler feeds by poking its head above ground at night and reaching all around its burrow to grab leaves with its toothless, padlike mouthparts. It can be a picky eater. Maple leaves apparently taste better than oak leaves, for instance, and may disappear faster.

Darwin spent enormous energy demonstrating that worms think their way around a leaf almost as if it were a geometric problem, usually seizing the narrowest point to reduce resistance as they drag it down into the burrow. Subsequent researchers have suggested that worms can also learn their way around a simple maze. The rest of the world, however, has not rushed to embrace Darwin's argument that the spark of active intelligence exists even in a worm. (In 1881, *Punch,* the humor magazine, took Darwin's whole theory as an affront: Mixing earthworms and evolution, one of its cartoons traced the ancestry of a British gentleman back to literal vermin. "Man is but a worm," the caption pined.)

Whatever the worm eats, bits of sand or rock in the gizzard are rolled round by the powerful muscles of the digestive tract to work like millstones in breaking down the meal. The nightcrawler keeps its hind end tucked securely in its burrow while it feeds, for rapid retreat. If a robin grabs the worm, it can expand its lower segments to wedge itself in the burrow. The worm's skin is also lined with retractable bristles, or setae, normally used for locomotion, and it can drive these into the walls of the burrow as an anchor. A worm may snap in two before its antagonist can yank it out. Contrary to folklore, however, both halves don't regenerate; a worm can replace a limited number of front or hind segments, but only if it's still got a middle.

Animals fare better than humans at turning the worm to profit. Foxes, for instance, learn to hunt worms in the leeward side of copses, out of the wind, especially on rainy nights. They crisscross a good area methodically, ears up, alert for the telltale sound of bristling setae scraping the soil. Then, writes mammalogist David Macdonald, the fox "plunges its snout into the grass, grabs the worm with its incisors and swiftly raises its head." Macdonald watched foxes snap up as many as ten worms a minute and reported that in some seasons worms may be a fox's main source of food. At times, worms may also be important for grizzly bears, which key in on areas where worms gather in wet weather. Badgers, moles, shrews, the tweezer-beaked rat in the Philippines, certain carnivorous slugs, and a host of other creatures prey primarily on worms.

No one knows why phenomena like rain and vibration bring out worms. One intriguing theory about predation on worms is that gulls and certain turtles may actually paddle the ground with their feet to bring worms to the surface. A researcher in central Pennsylvania observed the improbable sight of a wood turtle stomping its front feet and banging its plastron down on a moist stream bank, then moseying around to gather up the outraged worms. Perhaps the same effect, magnified to the scale of a Boeing 747, played a role in one shining moment of extraterrestrial earthworm glory: Earthworms were implicated as an aviation hazard at New Zealand's Christchurch Airport because they were keeping so many gulls fat and happy along runways, where the gulls tended to become turbine fodder.

To satisfy our own inexplicably widespread and remorseless urge to get worms out of the ground, we humans will resort to vibration, electrocution, inundation, chemical attack, or almost any other remotely plausible method. A school in the English town of Nantwich actually sponsors an annual competition in

worm charming, limiting entrants to some of the more sporting techniques. On a Saturday afternoon in June, they assault their three-square-meter patches of the Sceptered Isle with drum, clarinet, seesaw, a tap-dancing pony, or almost any other device. In 1980 one contestant set the world worm-charming record of 511 worms in thirty minutes by the conventional method of jabbing a pitchfork into the ground and twanging it. The winner gets a golden worm rampant.

In North America, we tend to be more earnest about our earthworms, possibly to our own detriment. Many people use electric probes to jolt fish bait to the surface. The U.S. Consumer Product Safety Commission has reported at least twenty-eight deaths by electrocution involving worm probes. In the Apalachicola National Forest, investigators suspect that worm grunters sometimes start their own forest fires to open up new territory. One such fire burned seven thousand acres after Memorial Day weekend in 1992, when the price of worms leaped to an all-time high. Forest rangers say fighting the conflagration cost them about $400,000.

But the Canadian worm business makes Florida's grunters look, astonishingly, like hayseeds. Worm gatherers in Ontario sometimes pay thousands of dollars for seasonal harvesting rights to orchards and golf courses, and they export a billion worms a year to the U.S. bait market. Pickers work at night, stooped over, with an open tin can strapped to each ankle, and a miner's light for spotting their quarry. Since the worms come to the surface to feed anyway, without artificial inducement, a good picker can gather ten thousand worms in a night. Until the business became more professional in the late 1980s, bait traders used to assemble for breakfast at a restaurant in west Toronto, to bid in a kind of crawler commodity exchange.

The tolerant reader who has stayed with me this far may begin to sense that the earthworm isn't merely nature's drudge, but can also sometimes slip out the side door into the blue surreal. (Did I mention that Darwin played the piano to his worms when he was trying to sort out the effects of sound versus vibration?) With earthworms, you begin to imagine that *anything* is possible. This may partly explain why so many people end up putting their last, best hopes into the strange proposition that whatever the question worms are the answer.

In something like this optimistic spirit, I phoned one of those "healthy planet" mail order companies not long ago and ordered my own personal worm bin. The catalogue showed a Martha Stewart type dropping a spotless banana skin into a plastic box. Worms "in a closet or under a table" would eat my family's kitchen waste "without any mess or bad odors" and turn it into rich potting soil. "And don't worry . . . the worms can't escape!" I read that part aloud to my wife. No slimy stampedes to spoil the big dinner party. Bring on the vermicelli!

The small-is-beautiful dream of becoming a worm wrangler and a master of compost first gained currency in the 1970s, and it's one of those neglected life-style choices that seem to come back and haunt certain people in midcareer. In those days, researchers produced promising evidence that worms could ease the problems of municipal-garbage and sewage-sludge disposal. They could turn manure from intensive pig and poultry operations into salable compost—and then the farmer could harvest the worms themselves for their protein. Several studies showed that chickens gained weight faster on earthworm meal than on a normal diet. Researchers in the Philippines found that homegrown earthworms were a good alternative to imported feeds for rearing quail and tilapia. A Canadian provincial government looked into schemes to raise worms for dog food, among other products.

The current, persistent rumor that worms are a secret ingredient in fast-food hamburgers may derive from these efforts (to which the quality of fast food also adds a frisson of plausibility). Earthworms have, in fact, served as human food for the Maori of New Zealand and as an ingredient in a Japanese pie. California State Polytechnic University in Pomona actually sponsored an earthworm cooking contest, and the finalists included "Earthworm Patties Supreme," "Curried Ver de Terre and Pea Souffle," and "Applesauce Surprise Cake." Though worms are as much as 90 percent water, their dry weight is comparable in protein content to almost any other meat. The trouble with the fast-food rumor—and with almost all worm-ranching schemes—is that it would be cheaper to buy tenderloin.

Alan Tomlin, a researcher with Agriculture Canada, keeps a file on failed attempts at the earthworm game. Promoters often advertise in areas experiencing hard times, he says, where desperation helps give the pitch a tantalizing, perpetual-motion appeal. Buyers may pay $4,000 to $7,000 for a franchise, which gets them worms, several dozen plastic buckets, a "secret technology" manual, and the "exclusive" right to sell worm franchises to other suckers. Then they find that growing worms is costly, time-consuming work, with no profitable market for the product. Tomlin estimates that rearing worms can cost a small operator $30 a pound.

My own worm bin, purchased not for pocket money but for the greater good of the planet, arrived when I was on an out-of-town business trip. By telephone, I explained to my wife, Karen, how to tear up old newspapers for bedding and make a gloppy knot of a thousand compost worms (*Eisenia foetida*) feel generally at home in our kitchen. And what a happy talk we had!

It slipped out that this worm scheme had already cost $92. I explained that if we removed *just five pounds* of garbage a week from the waste stream, the town would save enough in dumping fees to justify the cost of the worm bin in no more than—what?—13.6 years. Karen pointed out that $92 was the price of dinner for two at a fine restaurant—an extra cost I was going to have to factor into the equation at the earliest possible date.

When I wandered home two weeks later, the worms were churning their way through a sea of newsprint and grapefruit rinds. I peeled back a layer of garbage, and they reared their ugly little heads and went wriggling for cover. "Could there be more of them in there than when we started?" Karen demanded. I noted learnedly that worms are hermaphrodites, equipped with both male and female organs. When they mate, they exchange sperm for external fertilization of the eggs in a kind of mucosal ring around the swollen part of the torso. Some also reproduce asexually by parthenogenesis. Karen gave me a look suggesting that what two (or fewer) consenting adult worms do in the privacy of their own bin is their business. So I cut to the bottom line: In one study, eight compost worms multiplied to fifteen hundred individuals in six months. "Oh God! What are we going to do?" she said, with a lovely tremolo bordering on suppressed sob. "Sell them for fun and profit?" I didn't tell her that compost worms make lousy fish bait or that I was having certain squeamish thoughts about becoming a fly fisherman.

A few weeks later, the kitchen started to smell like a compost heap. Then we had a breakout (a "worm crawl" in the jargon), and though the dog happily lapped up the escapees, Karen summarily banished the whole enterprise to the root cellar.

But none of this is intended to discourage proponents of the earthworm (unless, that is, they are thinking of dropping $7,000 on a major career change). Worms can indeed accomplish miracles. Dutch researchers are using them to help reclaim freshly drained lowlands, and worms seem to serve as a natural plow in the promising field of no-till agriculture. Worms don't merely plow the Earth, as Darwin observed a century ago; they may also produce plant-growth hormones. Studies have shown that active worm populations dramatically boost plant productivity.

Even as a solution to the solid-waste problem, worms may yet prove to be more than a pipe dream. North of San Diego, the Fallbrook Sanitary District until recently ran this country's most successful large-scale worm composting scheme. Serving seventeen thousand people, the district produced ten tons of wet sludge daily. A regimen of regular composting for thirty days, followed by sixty days' treatment by compost worms turned the sludge into a premium garden product. Local nurseries found that they could sell so-called "Fallbrook Vermigro" for about $5 a bag. The program was actually showing a profit, counting the avoided cost of trucking the sludge to a landfill. Then neighboring homeowners decided that all composting smells bad. After a bitter fight, the uncomposted sludge is now being trucked 100 miles out to the desert, and the worms have been sold.

The city of Seattle has taken an alternative approach to vermicomposting. Their program was inspired by the nation's reigning worm guru, Mary Appelhof, a designer of worm bins and the author of *Worms Eat My Garbage* and its ambitious sequel, *Worms Eat Our Garbage*. Appelhof is a delightful, easygoing Michigan resident who sometimes wears a "Worm Woman" hard hat. Seattle is encouraging residents to build their own worm bins, which can be camouflaged as patio benches.

"I've been working with earthworms for 20 years," says Appelhof. "I used to dream eventually we would get tons of worms eating tons of garbage. But I didn't have the means to make that happen. Now Seattle has 6,000 worm bins, and say each is handling 5 pounds of waste a week, which is 15 tons, which is 780 tons a year. So tons of worms *are* out there eating tons of garbage. It's happening. I just didn't know it was going to happen on the individual scale."

It is a highly likable dream, with a one-step-at-a-time logic, aptly suited to Darwin's idea that small causes yield great effects. It certainly makes more sense than trucking tons of solid waste out to vanishing landfills.

But as Darwin also discovered, the closer you look at earthworms, the more complicated they get. Add people to the equation, and you have to wonder how well worm bins will go down with the average self-indulgent American—such as myself. Just now, I am sitting in the root cellar flipping through Appelhof's first book. My worm bin is humming along nicely. But I see that sooner or later I'm going to have to separate several thousand worms from the rich compost they're producing and start over. Appelhof suggests several methods. The one that catches my eye involves three hours of "therapeutic" hand-sorting. It's due about two or three months after start-up.

I haven't told Karen yet, but the way I figure it, that'll be just about time for my next big trip out of town.

A Small Point of Interest

Just now I was reading one of my old journals from a trip somewhere in South America, when I turned the page. There, flattened next to the binding, was a dark smudge of mosquito and, on the opposite page, its Rorschach image in dried blood, probably my own. In that instant, all the unheralded charms of the rain forest came rushing back to me: the way my clothes were always caked and sodden with mud, the way the howler monkeys roared their jocund welcome and flung sticks at my head, the feeling of sliding down wet clay trails and over a wobbly one-log bridge at midnight, in the endless rain, with a dehydration headache welling up behind my catatonic brow. But above all, I recalled the relief of coming back to camp, to sleep and give sustenance to mosquitoes.

For those of you who have not experienced this almost sacramental moment of union with Gaia, here is what it feels like: You drop your clothes in a clay-heavy heap, leaving your puckered flesh bare just long enough for the mosquitoes to roar in like sooners at a land rush. Then you dive into your individual cocoon of muslin drapery, and spend the next ten minutes slapping and spattering winged droplets of your own blood all over the netting and the sheets. Then, having killed the last mosquito, you recollect that you have forgotten to pee, climb back out (after longing, briefly, for a catheter), and do it all over again.

When the sound of slapping finally dies away, there is a moment in which the hostility and bone weariness of the day succumb to peace. Someone looks up bleakly at his white shroud and remarks: "I feel like a pupa." From under his netting, someone replies, "I was beginning to think of you as a maggot." And someone else wonders aloud if they might all have a better attitude had they brought along one of those soothing tapes with names like "Rain Forest Retreat": "A myriad of beautiful butterflies dance in the shafts of sunlight that filter through a warm, soft mist . . ." Suddenly sleep knocks you on the head like a hammer. You flop sideways against the drapery, and the mosquitoes instantly congregate on the other side to thread their proboscises through the muslin, penetrate your flesh, and suck you dry.

It could of course be worse.

In the Arctic tundra, for instance, you could get frozen and sucked dry at the same time. The spring snow melt hatches all the dormant mosquito eggs at the same time. When the adult mosquitoes emerge en masse they have about twenty minutes to mate, find a victim, get a blood meal, and lay a new batch of eggs before winter sets in again. Canadian researchers once sat still in such a swarm long enough to report that they suffered nine thousand bites a minute, a rate sufficient, at least in theory, to drain half their blood supply in two hours.

We are lucky by comparison. Apart from the occasional ill-considered expedition, we live in the vast middle zone, neither arctic nor equatorial. We get to nurture our heightened environmental feelings from behind screen windows, and blithely forget how much we owe to the depredations of public health officers. It is comforting to forget, for instance, that before the marsh drainage programs of the early twentieth century, the suburbs north of New York City were a malaria zone, or that mosquito-borne yellow fever once killed 10 percent of the people in Philadelphia.

It is pleasant to forget that at the Jersey Shore, where I frolicked in my DDT-misted youth, my grandparents once spent the best days of the season stewing and slapping on the front porch, citronella to one side, lighted punks to the other, and a barricade of smoke fires banked up on the lawn before them. The "Jersey terror" on occasion rose from the salt marsh in dense clouds to torture visitors and ruin otherwise fine hotels. At the height of one season in Atlantic City, tourists maddened by the constant biting fought their way onto homebound trains and even bribed conductors to let them aboard. Other years, they masked themselves in handkerchiefs for protection, or plugged away hopefully at stinking cigars. Even well into this century, children going fishing used mosquitoes as an excuse to smear their faces first with axle grease.

In truth, apart from ourselves, few people have lived at any time without knowing the mosquito on a much too intense and personal basis. Among weak-hearted Pilgrims, the local "muskeetos" were good reason to flee Plymouth Rock, and according to one account, George Washington complained that a New York species bit him through the thickest boots. But being insufferably annoying isn't enough.

Mosquitoes, not king cobras, not grizzly bears, not great white sharks, are the most dangerous animals on Earth. They transmit the deadliest diseases known to man—malaria and yellow fever—as well as filariasis, dengue, encephalitis, and a hundred or so other maladies. We only began to figure this out a century ago, and we haven't quite gotten it into our heads yet. So fear of mosquitoes is not part of our genetic code, as fear of snakes or spiders seems to be. Hollywood does not make spine-tingling thrillers about the mosquito menace. Mosquitoes are just too small and familiar to seem like much of a threat. When government health officers tried to end a 1905 yellow fever epidemic in Panama, some local residents actually tore the screens off their windows and doors to show their contempt for the idea that they could be killed by a creature as trivial as the mosquito.

But the truth is that mosquitoes have affected every aspect of our lives, from the language we speak to the prominence of nations. With no greater logic than their own need for blood, they have ruined one man's sleep, killed another's family, ravaged armies, and broken empires. In 1741, when England sent 27,000 men to destroy Spanish power in the New World, a yellow fever epidemic killed 20,000 of them. When France sent 33,000 men to secure its control of the Mississippi Valley in 1802, yellow fever killed 29,000, leading directly to the Louisiana Purchase, in which the United States bought, at a fraction of its worth, the entire territory from the Mississippi to the Rocky Mountains.

The mosquito has wielded even greater influence with malaria, though in a slower, steadier, more insidious fashion. This disease, named for the noxious airs (literally *mal'aria* in Italian) long thought to cause it, does not usually kill its adult victims outright. Instead, it gradually weakens them, leaving them prey to other diseases. As one health consultant has put it, "whom malaria does not kill, it enslaves." As recently as the 1930s, the United States used to report 100,000 cases of the disease a year. But in the tropics, the caseload has always numbered in the hundreds of millions. The cumulative damage is such that, according to the U.S. Centers for Disease Control, malaria has probably had "a more profound influence on world development" than any other disease.

Having identified the mosquito as the source of so many troubles, we seemed, for a brief moment at mid-twentieth century, to be capable of escaping its dire influence. Malaria and yellow fever were eradicated in the United States. In India, heavy spraying with DDT reduced the number of malaria cases from about fifty million in the early 1950s to just fifty thousand by 1961. But beneath the spectacular public health triumph, what was really going on was a case of evolution by unnatural selection: High technology took out the relatively easy targets. But a few mosquitoes made of sterner stuff survived, thanks to some hitherto irrelevant quirk in their genetic makeup. Because invertebrates reproduce rapidly and because the habitat was suddenly wide open, these incorrigibles took over. With the ascendance of pesticide-resistant mosquitoes, the number of malaria cases in India shot back up to fifty million. In some areas, the disease organism carried by the mosquito, a protozoan of the genus *Plasmodium,* simultaneously evolved resistance to the most effective drug treatments for malaria. Today, some *Plasmodium* strains in Southeast Asia have actually evolved resistance to antimalarial drugs so new they have not yet been approved for use in the United States. If there can be such a thing as too much evolution, this is it. Would someone please send in the creationists?

Malaria now afflicts about 300 million people, most of them in Africa, and kills as many as three million people each year, most of them small children. In the 1990s, the disease has also reappeared in the United States, with locally acquired cases in California, Texas, New Jersey, and New York. Our moment of victory over the mosquito was perhaps merely an interlude between plagues.

Disease and nuisance alike begin with the common mosquito bite. There are more than thirty-four hundred mosquito species in the world. Some bite only in daylight, others only at dusk, still others only when it is dark. But the bite itself is essentially the same whether it occurs in London or Kuala Lumpur. It is an act worth contemplating, even as the hungry mosquito inserts her pointy tip and draws your fresh blood up into her belly.

Only the female mosquito will bite you; the male survives entirely on nectar. But the female needs protein to produce her eggs. So on a perfect summer night, when the air is just cool enough for you to sleep comfortably with the sheet drawn up over your shoulder, she enters your bedroom in search of blood.

She is a frail, translucent creature, this mosquito, yet she is amply equipped for her task. Her quarter-inch body is divided into three parts: a head no larger than a pin's head; a hunch-backed midsection, or thorax; and a segmented abdomen, longer than the first two parts combined. From those first two parts, a spiny array of wings, legs, and antennae pokes out in all directions. Taken as a whole, she seems an unlikely aircraft.

But the mosquito shows remarkable strength and agility in flight. So much so, writes J. D. Gillet, an English entomologist, in his book *The Mosquito,* that some species can fly through rain, dodging each drop, and arrive, still dry, at their destination. The mosquito can hover, loop-the-loop, speed up suddenly or slow down, dart in and out between slapping hands, even fly upside down, sideways, and backward.

Throughout, her wings beat from 250 to 600 strokes a second, depending on her species and, to a lesser extent, her speed. (By contrast, the bumblebee has a wingbeat of 130 strokes a second, the butterfly, as few as eight strokes a second.) She achieves such a rapid wingbeat not by continuous conscious exertion, but by switching on the unusual wing muscles in her midsection. Until she switches them off again, these muscles automatically contract and relax at a prodigious speed.

The thirsty mosquito has better things to think about: As she prowls your bedroom in search of a suitable victim, she samples the air with thin sensors feathered out along her two antennae and three pairs of legs. These sensors alert the mosquito to warmth, moisture, odors, carbon dioxide, some sounds, and possibly other stimuli. Day-biters, such as the salt marsh mosquito (*Aedes sollicitans*), have an easier time of it: They may spot you first with their huge eyes. The sensors take over only at close range. But biting by night is an excellent defensive strategy for obvious reasons. "Consider the outcome if you were to approach an elephant with a syringe," writes Marc J. Klowden, a University of Idaho researcher who characterizes himself as "a mosquito psychologist." Clearly, it is better for the mosquito if her victim is asleep, or, if awake, hampered by darkness from effective swatting.

The mosquito also cannot see in the dark, but her sensors allow her to pinpoint your location. She catches your scent by the stream of exhaled carbon dioxide downwind from where you are sleeping. As she flies randomly about the room, she enters this stream and continues straight ahead until she passes out its other side. Immediately, she turns back, reenters the stream, and, like a sailboat tacking into the wind, angles slightly in toward the stream's source: your unwary, exhaling self. Gradually, she zigzags her way to within a few inches of you. Now, for the first time, you may notice her by her infernal buzzing.

In a similar circumstance, D. H. Lawrence once wrote:

> *It is your trump,*
> *It is your hateful, little trump,*
> *You pointed fiend,*
> *Which shakes my sudden blood to hatred of you:*
> *It is your small, high hateful bugle in my ear . . .*

This unsettling alarm is caused by the mosquito's rapid wing-beat, the vibrating muscles of her midsection, and the fine scale-like structures over the respiratory openings on her thorax. The exact tone varies from species to species and plays a vital role in bringing male and female of the same species together for romance.

Indeed, the buzzing may have brought at least one adulterous human couple together as well. In the first century B.C., the Greek poet Meleager asked a mosquito to seek out his sleeping lover and "swiftly with a skimming kiss" whisper in her ear that she should come to him. "Up, you musical thing!" the poet urged the mosquito, "Up and away! To wing! To wing!" Then, as a parting thought, he reminded the mosquito to "say it low" lest it wake her husband. Clearly, D. H. Lawrence simply lacked a sufficiently amorous imagination.

Just now, as the mosquito circles noisily above you, close range sensors, attuned to moisture, warmth, and substances in the sweat, begin to work. With these, the mosquito determines whether you are a suitable victim. On close inspection, a given mosquito might find you unappealing, or may simply prefer your spouse. In a West Indies study, for example, mosquitoes favored the father in one family four times more often than his wife (this would have been bad luck for Meleager) and as much as 150 times more than their baby.

"You may go out with your wife," one mosquito specialist advises, "and she's getting bitten when they're not even bothering you. But when she's not there, you're it, you're elected. Each person has a different body chemistry; there are different substances on the skin. That's what it boils down to."

Assuming our mosquito finds you perfectly delectable, she now begins searching, still using her antennae, for a place to bite. Eccentricities abound. Gillet, for example, notes an African species that bites mainly on the scalp, even through thick hair. Once, in Peru, I saw a colleague come back into camp with ten mosquitoes on one elbow and another firmly embedded between his eyes. He was an unappetizing specimen to start with, but there is no point disputing taste in mosquitoes. They've been sucking blood long enough to know what they like (and better him than me). The mother of all mosquitoes probably started out lapping blood from the wounds inflicted during dinosaurian battles of the Jurassic period. (Hence Michael Crichton's premise in *Jurassic Park* that dinosaur DNA could survive in the bellies of mosquitoes preserved in amber.) From wound-feeding, early mosquitoes rapidly evolved the means to steal blood covertly by penetrating the skin.

After a moment, the mosquito settles down, ever so gently, on your flesh. You will probably not notice this. Mosquitoes are extraordinarily light-footed. One species can land on a spider's web without alerting the spider. Nor are you likely to notice as the mosquito deftly pierces your skin with her pointy weapon.

This proboscis, jutting from just below the mosquito's eyes, is no mere hypodermic needle. It contains six sharp instruments, called stylets, all finer than hairs: two tubes, two lancets, and two serrated knives. A gutter-shaped sheath wraps protectively around these stylets along their entire length. Finally, a clamp attached to the sheath at the tip of the proboscis cinches the six stylets together in a single bundle. With this formidable instrument, the mosquito performs a masterpiece of phlebotomy (from the Greek for "vein cutting").

She lines up on your skin. The maxillary palps, two stubby, segmented appendages sticking out just above the proboscis, suddenly lift at a seventy-five-degree angle, like a conductor's baton at the start of a symphony. Then the mosquito drives all six stylets into the skin at the same time. Finding the sweet spot typically takes her about fifty seconds, and you may notice the

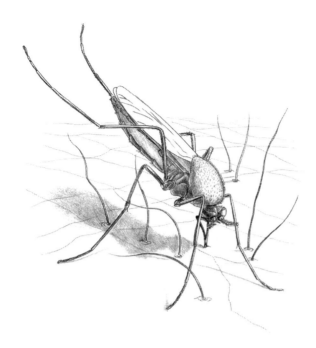

maxillary palps rocking slightly as she saws her way in with her serrated knives.

The stylets penetrate half a millimeter, about the length of a second-day whisker. The clamp on the gutter-shaped sheath remains at the point of entry as a guide through which the stylets pass. The sheath itself, being permanently connected only to that clamp and the mosquito's face, slips away from the stylets under the pressure applied by the mosquito and elbows neatly back out of the way. It will return to its normal position only when the mosquito finally withdraws the stylets and retreats.

But now, beneath your skin, the bundle of stylets is just entering the closely knit web of capillaries—veins themselves thinner than hairs. The mosquito seeks to penetrate or, more likely, to rupture one of these veins. If the stylets actually enter a vein, they bend to its shape and probe a fraction of a millimeter into it. The mosquito can hope for nothing better. She has tapped a gusher and will require only a minute to take a full meal from its steady flow. By contrast, she may waste three minutes, and risk her life, drawing up the leaked blood from a vein she has merely ruptured. Moreover, she may have to rupture the vein again before it will provide her with a sufficient meal of blood.

All this normally passes unsuspected by the victim. But just before the mosquito begins pumping up your blood, she dribbles a small quantity of saliva down the smaller of the two stylet tubes. Beneath your skin, this mixes with your blood and prevents it from clotting midway up the mosquito's proboscis. With the help of this anticoagulant, she can pump enough blood up the larger stylet tube to turn her long abdomen into a swollen, red paunch. Stretch receptors in her abdomen tell her when to quit. She can consume about four times her own weight in blood; but, bad as this may look, it would take fifteen to twenty such meals to equal a single drop from an eyedropper. Unfortunately, a residue of saliva remains beneath the skin after each bite to vex you with an itching welt. By then, more often than not, the mosquito has lumbered off, leaving her victim to slap and scratch in vain.

The mosquito now seeks only rest. Weighed down with your blood, she settles on a wall or nearby leaf to wait. During the quiet days that follow, her body will turn that droplet of blood into a hundred or more tiny, identical eggs.

Even before the mosquito has taken her first blood meal, she has mated, attracting a male of her species by the distinctive buzz of her wingbeat. That single mating suffices for the four or five times she will lay eggs during the month or two that she lives. Each time, as she prepares to lay her eggs, she fertilizes them with the sperm stored within her. Then she seeks water.

More than two thousand three hundred years ago, Aristotle suggested that the mosquito begins its life through spontaneous generation in water. He was half right. All mosquitoes pass through their immature stages—egg, larva, and pupa—in or around water. But spontaneous generation would have been too simple. The mosquito's breeding habits vary endlessly among the three thousand four hundred species. Some lay their eggs singly on the water, some in delicate, floating rafts of more than one hundred eggs, some out of the water entirely. Moreover, different species will find suitable water and breed prolifically almost anywhere: in tree hollows, empty snail shells, or the tiny bowl formed on a toadstool cap, in a clogged roof gutter, a cesspool, an old tire swing, even in water polluted by acid waste at the bottom of a coal mine. Virtually any quantity or quality of water will do for some species. One species (*Wyeomyia smithii*) has actually evolved to lay its eggs in the small pool of rainwater stored by the leaves of the carnivorous purple pitcher plant—a pool that otherwise serves as a death trap for insects.

While all mosquito eggs need water to hatch, some will hatch only after freezing through a winter, others only after drying out. Some eggs laid by the millions at floodwater levels may survive, like time bombs, three to five years on dry land until flood waters come again to hatch them. This is one reason heavy rains produce such a large crop of mosquitoes.

Typically, the egg hatches just a few days after being laid. The larva is an aquatic creature bearing little resemblance to the adult mosquito it will become. Though it has a head with eyes and antennae, its tube-shaped body lacks legs, wings, and the prominent mouthparts of an adult. The larva spends its brief life just below the water surface, breathing through a snorkel-like tube that penetrates to the air. To sustain its rapid growth, it feeds continuously on microscopic plants and animals. The pitcher-plant mosquito, for example, doesn't merely use its host plant for a reliable water supply in which to lay eggs. Its larvae also feed on the detritus of insects trapped and drowned there. The adult female mosquito actually chooses to lay her eggs only in young pitchers, because they are the most efficient insect-killers. Thus there should be plenty of leftovers for her offspring to eat.

The larva usually develops for four to ten days. Then, a split opens down its back through which the pupa emerges. This final aquatic form resembles neither the adult mosquito nor the larva even slightly; it is shaped like a comma and lacks even the characteristic antennae. It develops for about three days, breathing at the surface but not eating. Then, it too splits, along the top of its head. Just beneath the split, the adult mosquito—in its familiar form—now begins swallowing air.

Slowly, the mosquito's midsection and head expand and push up out of the pupa. Then it draws its abdomen, legs, wings, and proboscis up onto the surface. There, on the floating husk of the pupa, it rests, while its wings dry and its skin hardens. Finally, it flies. Over the next few days, its wings will achieve their distinctive beat and buzz, it will mate, and then, if it is female, it will seek the blood meal.

She will not necessarily search out a human victim. Some species may prefer elephants, some mice. One common North American mosquito (*Aedes canadensis*) bites turtles not just on the head, neck, and legs but even between the scutes of the shell. Wayne Crans, a Rutgers University researcher, once netted 569 such mosquitoes on a single, pitiable box turtle. Other mosquitoes don't bite at all. One tropical species preys only on a hardworking ant, drawing from its very mouth not blood, but honeydew. In a few species, male and female alike survive solely on the nectar of flowers and may play an important role in pollination.

But honestly, I have no heart for propounding healthy attitudes toward the mosquito. We are already too complacent. I'd rather dwell on the early explorers in Alaska driven by mosquitoes "to the verge of suicide and insanity" or on indigenous Brazilians periodically forced to bury themselves to the neck for protection. I'd rather think about the ten thousand mosquito eggs measured in a single square foot of marsh soil, and of all those newly emerged little vampires taking wing in one vast, sky-darkening, bloodsucking swarm, their collective need for a meal so sudden and intense that unlucky victims die of exsanguination, literally sucked to death.

It has happened in this country, though not to humans. In 1963 on the Louisiana coast, heavy rains brought off waves of mosquitoes from eggs left behind in previous flood years. Department of agriculture researchers found that the average steer was suffering up to two hundred mosquito bites a minute. By midsummer, mosquitoes had killed five hundred calves and two hundred adults by allergic shock or simple loss of blood. Also by suffocation. The mosquitoes came in clouds. Even the steers that survived weighed ten pounds less than normal by market time. When it happened again in Brazoria, Texas, south of Houston, in 1980, veterinarians performed autopsies and found that half the normal seven to nine gallons of blood in the animals' bodies was missing, apparently removed at the excruciating rate of one-millionth of a gallon per bite.

But I see that I am verging on hysterical entomophobia here. In truth, the likelihood that you or anyone you know will be sucked to death by mosquitoes is remote, probably on the order of the odds that you will die by a direct lightning strike during a shark attack on the shores of Lake Okeechobee. And yet . . .

Mosquito-borne disease is a genuine and worsening threat. Yellow fever, which once terrorized the entire world, is again on the loose in South America. An outbreak in Peru in 1995 claimed 350 victims, and half of them died. Dengue fever, which, like yellow fever, seemed to have been eradicated from the Western Hemisphere, is now epidemic in Latin America. In 1995, dengue appeared in Houston and in southern Texas. A more virulent form, called dengue hemorrhagic fever, broke out in Reynoso, Mexico, on the Rio Grande border. It causes internal bleeding, coma, and shock, and about 10 percent of its victims die. Encephalitis, which can kill or cause permanent brain damage, also remains active in the United States. Malaria, finally, is on the comeback everywhere.

But the one disease that seems to have focused public attention on the mosquito, paradoxically, is AIDS. The thinking goes that if syringes contaminated with blood can spread this dreaded virus, why don't mosquitoes? The answer has to do with the two routes, biological or mechanical, by which a mosquito can cause disease. Having bitten an infected animal, a mosquito may become a breeding ground for the disease organism, which ultimately spreads to its salivary glands and gets injected with the saliva into each new victim. This biological route is how the mosquito transmits malaria, yellow fever, dengue, encephalitis, and other viral diseases. In a laboratory study, the AIDS virus has actually survived for three days in a mosquito's gut. But unlike other viruses it does not reproduce or enter the mosquito's bloodstream. It gets digested, making the biological route a dead end.

Mechanical infection sounds, at first, more worrisome. When a mosquito gets interrupted in mid-bite on an infected host, some of this first host's blood remains in the mosquito's proboscis. It may then get reinjected when the mosquito begins to feed on its next victim. Research suggests mechanical transmission is possible with hepatitis B, a liver disease that, like AIDS, is commonly spread by blood exchange or sexual contact. But the hepatitis B virus occurs at a million times the concentration of the AIDS virus. Because the AIDS virus occurs at such low levels and because it takes a lot of virus to cause an infection, the fraction of a droplet of blood in a mosquito's proboscis poses no plausible threat. One official at the Centers for Disease Control has calculated that, to face any real risk, you would have to be bitten by two thousand eight hundred mosquitoes, each of which had been feeding immediately beforehand on an AIDS virus carrier. You stand a better chance of dying by exsanguination.

But if mosquito-borne AIDS is a nonissue, the virus has nonetheless thoroughly punctured our illusion of immunity from epidemics. It has become conventional wisdom that our overpopulated, jet-linked world is on the edge of some devastating new plague. Hence the current concern about exotic new pathogens and hot zones and outbreaks. Fickle public attention has yet to dwell on those old standbys—dengue, encephalitis, malaria, and the forgotten terror of yellow fever. But the unglamorous mosquito may still have its day.

In 1985, a new mosquito arrived in Houston as a stowaway in used tires shipped from Japan for the American retread market. The Asian tiger mosquito, *Aedes albopictus,* has spread rapidly through the American South, where it is already the most common biting mosquito in many cities. It has also established itself in Chicago, where it readily tolerates the frigid winters. The reason this matters is that *A. albopictus* can transmit dengue, encephalitis, and—in laboratory experiments, though not so far in nature—yellow fever. Like its cousin *Aedes aegypti,* which is the primary vector for yellow fever, it breeds prolifically in and around homes.

Scientists disagree about how serious a threat the Asian tiger mosquito may pose. While it is a more aggressive biter, it doesn't focus exclusively on humans, as the yellow fever mosquito does. Moreover, the Asian tiger mosquito has not yet caused any known disease among humans in the United States. The Centers for Disease Control has taken a conservative stance, advocating careful monitoring but no large-scale attempt at eradication. Some academic researchers and local health officials are more alarmed. One has likened the spread of the Asian tiger mosquito to a wholesale distribution of empty guns. "Sooner or later," he adds, "Mother Nature is going to pass out the ammunition."

The prospect of mass sickness or slaughter by mosquitoes in our own backyards may seem remote. But it happened as recently as 1905, within the lifetime of Americans still living today, when the last yellow fever epidemic ravaged New Orleans. Philadelphia, which suffered the disease more often than any other American city, had its last epidemic in 1902. During one of Philadelphia's epidemics, in 1793, the disease sent thousands of residents, including President Washington, into flight. The national government was in turmoil. Whole families died horribly. Bodies lay in the streets. Cemeteries were filled to overflowing. People lit bonfires to dispel the poisonous vapors they thought caused the disease—and still they died. Philip Freneau, the contemporary poet, lamented:

> Nature's poisons here collected
> Water, earth, and air infected—
> O, what pity,
> Such a city
> Was in such a place erected!

By the time the epidemic finally broke with the end of the mosquito season late in October, it had infected seventeen thousand people and killed five thousand. To equal that level of devastation, an epidemic striking modern Philadelphia would have to kill 150,000 people in just two months.

Well, it can't happen here. We know too much now, don't we? Still, as you lie in bed and wait for the next mosquito to come zigzagging in, and the next one after that, and so on through the summer, it may give you something to think about: Maybe, after all, we should be spending our precious tax dollars on obscure entomological research. Maybe we should be investing in some of the more thoughtful and environmentally sound methods of mosquito control. Maybe—O blasphemous liberal thought!—we should actually be spending more of our hard-earned dollars on foreign aid programs to help people afflicted by mosquitoes in less fortunate parts of the world. Maybe . . .

Naaah. Pull the sheet over your head and forget it. Get some sleep. Don't ask for whom the mosquito buzzes. Someone might tell you it buzzes for thee.

Ghosts on Wings

Once, not so long ago, people believed that moths were the souls of the dead flinging themselves against the windows of the living. "'Twas Annie's soul that beat outside and 'Open! open! open!' cried," a sentimental poet wrote, of a moth. Lonely travelers out on the moors encountered moths of a European species fluttering together at dusk and called them "ghost moths." Scientists named the genus *Hepialus,* which means "shivering nightmare."

So the story of an English clergyman who reared moths for a hobby fits the tradition. He wasn't a superstitious man. The crypt of the cathedral where he was canon was simply a good place to keep moth larvae over the winter. The oak coffin of a local duke, merely a convenient shelf. But one spring the canon arrived late to retrieve his box of larvae, and, under it, he found a series of neat holes drilled in the lid of the duke's coffin. Then he heard a rustling sound from within the coffin followed by something rap-rap-rapping. What the canon did next may seem odd. Alone in the crypt, he unscrewed the lid of the coffin.

It had dawned on him that the caterpillar of this species, sometimes called the carpenter moth, can bore holes even into an oak coffin, and, naturally, he didn't want to lose his specimens. Opening the coffin, he plucked the moths from the duke's skull. A true moth enthusiast must be prepared for the unusual, because the private lives of the moths are nothing, if not strange.

There's a sickle moon over Bear Mountain to the west, and the night sky is hazy and crystalline, like sugared water. The soft *pppt-pppt* is the sound of bugs meeting windshield as the station wagon races over the wooded Connecticut hills. "Lots of insects in the headlights, that's a good sign," David Wagner remarks, pushing the accelerator a little harder.

It's a perfect night for mothing, and Wagner already has his generators and black lights humming at strategic locations among the local lakes and fens. He is a zealous moth-man. The Gary Larson "Far Side" cartoon on the door of his office at the University of Connecticut depicts an "entomological rodeo," and you can just imagine Wagner launching himself out of the chute, hell-bent on moths.

At the first stop, he squats down in the pale violet gloaming of the black light. He sweeps the beam of his flashlight across the white sheet he has hung up as a backdrop, which is seething with long-horned caddis flies, stag beetles, giant water bugs, and two hundred or so species of moth. Wagner picks out a rosy maple moth the colors of a peach melba, and a brilliant green-and-black mottled leuconycta. He points out a chrome-plated creature smaller than the crescent on his fingernail. "You realize how many of these tiny things are moths?" he asks, pushing the Boston Red Sox cap back off his high forehead. "It's alarming. For me. I have to collect 'em.

"The pitcher plant moth. Book 'em, Danno," Wagner tells a helper. He spots a small moth perched on this reporter's knee, wings folded, with a profile like a steam locomotive. "*Autographa ampla!*" he says, and scoops it into the killing jar, which has a cyanide-laced rag at the bottom. The specimen will go into the university collection, as part of the supporting evidence for this inventory of local moths.

After a while, Wagner steps back to survey his winged riches, which seem actually to have increased in number, *thickened in density,* with the waxing of the summer night. Someone points out a *Polia* moth. "Where's *Polia?*" he asks. "It must be getting toward midnight. I'm not getting enough done. Oh, man, I'm just going to have to pick sixty species that I want to collect." The moths are now flying down our shirts, foundering in our scalps, flapping blearily across our mouths, and otherwise threatening to overwhelm us.

"Look at that abdomen," Wagner comments, picking a specimen off the sheet. "Purple, orange . . . ooh, he's oozing poison on me! He's bleeding right out the thorax, on purpose. Anybody want to taste it? All I can tell you is that it's bitter." But by now everybody seems to have eaten their fill of moths accidentally.

According to a familiar biological saying, a close study of nature suggests that God suffered from "an inordinate fondness for beetles." But clearly He liked moths, too. There are more than 150,000 species of Lepidoptera in the world (15,000 of them in North America), and almost all of them are moths. There are more species of moth than all the mammals, birds, fish, and reptiles put together.

You surely would not guess this from the popular literature. Butterflies, those flashy arrivistes, get all the publicity, though they constitute less than 7 percent of the Lepidoptera. They get their picture on postage stamps. They get mawkish song lyrics ("Butterflies are free to fly, fly away . . ."). They get ecotourism and butterfly zoos.

Moths, meanwhile, get frass, or possibly much worse. The hugely popular novel *Silence of the Lambs,* for example, features a psychopath who breeds death's-head moths, murders women gruesomely, and plants a pupa in each of his victims' throats.* Moth-wise, the book has a redeeming feature: The FBI agent heroine ultimately finds love with a normal sort of moth-man, a lepidopterist at the Smithsonian. But while the first moth-man made it into the hit movie version, the second didn't, and this is how it goes with moths. The last time they got good media buzz was when Mothra wrapped Godzilla in silk.

Moths have a reputation for being drab, brownish, and boring. Also bad: Some of our most desperate field and garden pests are leaf-ravaging caterpillars, like the corn earworm and the tobacco budworm, which grow up to be moths.

But moths are important pollinators, too, and many plants would go extinct without them. As caterpillars, they are the primary food of songbirds. (In fact, moths get eaten by everything from shrews to the grizzly bears at Yellowstone National Park, which gorge themselves on army cutworm moths to fatten up for hibernation.) They are also the source of the five thousand-year-old silk trade. Moth silkworms were once so highly esteemed that, in one of the earliest recorded acts of industrial espionage, priests hid them in the hollows of their canes and smuggled them out of China.

---

*Butterfly enthusiasts don't, in truth, come off all that much better. In one of his novels, Joseph Conrad portrayed a lepidopterist stepping obliviously across three fresh human corpses to collect a butterfly—a prize so lovely it caused him to sit down amid the carnage to contemplate the sweetness of life.

"I started out collecting butterflies. Then I grew up and switched to moths," one of Wagner's helpers, a collector named Ben Williams, remarked. "People think they're all ugly miller moth kinds of things, and for that reason they became more fascinating to me—for the variety of size and shape and color and life history. The biggest Lepidoptera is a moth. (It's a Southeast Asian species with a twelve-inch wingspan.) The smallest Lepidoptera is a moth. The ugliest Lepidoptera is a moth."

Williams, a retired prep school headmaster, was once a Marine captain protecting the President's retreat at Camp David, where some of his men noticed that he took an inordinate interest in life around the mercury vapor searchlights. "They'd say, 'Sir, I have a moth you might be interested in.' They always squashed it with the butt of their M-1s and they'd pull this mashed beast out of a cartridge belt." So Williams doesn't even bother to mention that the most beautiful Lepidoptera are also moths. But at home he keeps drawer after drawer of the showier moths, each specimen meticulously preserved: bat-size cecropias and sherbert-colored saturniids, moths with tiger stripes and leopard spots, hawkmoths that rival hummingbirds in size, shape, and color. "I'm a sucker for a pretty face," Williams confessed, the day he showed me his collection.

Moreover, the lives of the moths are far more curious than simpleminded, mothball-flinging humans generally imagine. Some moths, for example, practice virgin birth. In other species, the females are sexually promiscuous and seek out toxic males as a way to improve the health of their offspring. One moth caterpillar eats only antlers, another moth pierces grapes for their juice. Yet another survives on a diet of tears, fanning out below a cow's eye like facial armor. One moth species, in Southeast Asia, can pierce skin and suck blood. One spits cyanide. A moth caterpillar that burrows in legumes for its food is the reason the Mexican jumping bean jumps. (But you are probably still thinking about those toxic males and sexually promiscuous females, so be assured, gentle reader, that we will come back to them in due course.)

Even among species that confine themselves to the more familiar moth business of sipping nectar from flowers, it often seems that there is no ecological niche too obscure for a moth to exploit it. In one famous example, Charles Darwin examined an orchid from Madagascar with a foot-long nectary tube and postulated that there had to be a hawkmoth with a tongue almost that long to pollinate it. The moth eventually turned up, forty years later. As it hovers in front of the flower, it scrunches up its head to force the blood out into its huge proboscis, which uncoils like one of those paper squawkers at a kid's birthday party. Then it inserts the tip of the proboscis into the nectary opening as if threading a needle. There are advantages for both flower and moth in evolving toward greater length. So researchers, having discovered an even larger orchid species, are now seeking a hawkmoth with a sixteen-inch tongue.

If you are getting the idea that nature does not know where to stop with moths, you are correct. She is not content, for example, merely to give us the quintessentially indolent sloth, hanging plantlike in the rain forest treetops. She must also one-up herself, giving us the even idler sloth moth, which breaks off its wings and idles its life away in the sloth's fur. When the sloth clambers down a tree trunk for its weekly defecation, the female moth drops off to lay her eggs in the fresh dung, on which her offspring will wallow their way toward adulthood.

It begins to make butterflies, for all their flash, look just a tad shallow. There is a moth for every season, and the tantalizing thing is that they exhibit their startling variety not just in some South American cloud forest, but in the average suburban backyard, as we sleep. Wagner has identified more than eight hundred different moth species at the black light behind his

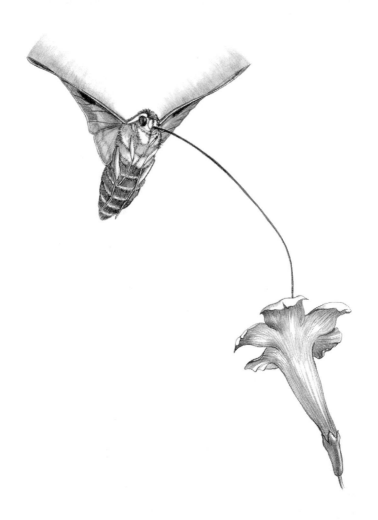

Connecticut home. He specializes in microlepidoptera, or what Williams refers to as "itty bitty little schnitzers." During the busy season, from April to October, he turns up a new one, unknown to science, every week or so. To a newcomer like me, even the commonplace moths can seem bizarre.

"Here's a really pretty bird-poop mimic," Wagner remarks, and there beneath the light perches a black-and-white splotch of a moth, cylindrical and with a moist-looking sheen. "They rest on leaves by day," he explains, "so it's a good thing to look like bird poop." So good, in fact, that thirty different kinds of moth just in the Northeast have evolved this disguise as a clever way to make hungry birds look elsewhere.

A moth fossil survives from the Jurassic period. But moth species really began to proliferate, together with the flowering plants, in a great evolutionary outpouring about 130 million years ago. The butterflies evolved from the moths at about the same time. Apart from the appalling difference in public relations, it isn't always easy to tell moths and butterflies apart.

As a rule, butterflies fly by day, depending on bright colors to tell their archenemies, the birds, that they eat plant toxins and taste bad. They perch with their wings primly held up, tips together. Their antennae tend to have clublike knobs at the tips.

Moth antennae look more like feathers, and moths generally fold their wings down over the abdomen, like an old cloth coat, the better to vanish furtively into their surroundings. By day, they lie low disguised as dead leaves, lichen, peeling bits of bark, or almost anything else a hungry bird might overlook. They generally fly at night, to avoid birds.

But it is unwise to be too confident about any of this. Many moths fly by day. One hawkmoth goes through the usual stages of lepidoptera development—egg, caterpillar, and pupa—and then, on emergence, experiences a sort of secondary metamorphosis: It sheds the scales from its wings in a black puff and buzzes off on transparent wings, looking like a bumblebee. ("No butterfly would ever do that," Williams remarked, displaying a drawerful of faux hornets, yellow jackets, and bees.) Not only is the coloration of the abdomen perfect, but the moth will tuck its tail under as if to sting, startling any would-be predator just enough to make its escape.

This kind of cross-dressing occurs even in caterpillars: Wagner turned up a caterpillar in his backyard that starts out disguised as a bird dropping. Then it gets too big for the part and, with the help of huge eyespots on its back, pretends to be a snake. If this doesn't startle any bird peering into the caterpillar's daybed, says Wagner, "they have a backup—an extrusible structure loaded with noxious compounds that they wipe all over their adversary. One of the substances is butyric acid, which is a sickly sweet smell in vomit."

All of this is enough to make a moth enthusiast think differently about backyards: Maybe, in the great scheme of the universe, it isn't your yard at all? Maybe—but how to put this delicately?—it's really just a nocturnal playground for cross-dressing and chemically dependent moths.

Sooner or later the moth enthusiasts I spoke with all sent me on to Noel McFarland, who is the unequaled student of his own backyard. McFarland seldom leaves his five-acre Ash Canyon property in the Huachuca Mountains of Arizona. His wife and daughter share the traditional Western fondness for what he calls "big habitat-destroying mammals," meaning horses, but he prefers moths, and there are plenty of them inside his fence line: He's collected about a thousand species so far, "counting only macros. If you added micros you could easily double that."

McFarland started black-lighting as a boy at his family's house in Beverly Hills, where he collected almost four hundred moth species on the back porch. After graduate school, he moved to south Australia, and became so obsessed with the local looper moths that for six years he covered a tolerant landlady's dining room table with rearing cages and her kitchen with pupa pots. The result was a weighty book: *Portraits of South Australian Geometrid Moths.*

He's been in Ash Canyon since 1979, and the night I visited, he took me on a tour of the six black lights he operates on the property. At age fifty-eight, McFarland is a slightly built man in a floral shirt, with weathered skin, a close-cut beard, and a few short ringlets of gray hair down over his brow. He has a thin, reedy voice, and, as he stands at the black light, the contented air of a man utterly in his element, bathed in purple light and with moths clinging to the insides of his eyeglass lenses. "These butterfly idiots are beating their brains out running around in the sun. They have to carry their equipment with them," he said. "Meanwhile, we turn on the lights and the moths come to us. They just haven't got it figured out yet."*

By the back porch, an evening primrose, the food of a certain sphinx moth, was flowering in the moonlight. The world of the lepidopterist, McFarland was saying, has its own social structure: Society looks down on butterfly enthusiasts, who look down in turn on moth lovers. The micro moth people worry that the macro types have a trophy mentality; the macro people mutter back that the micro types are pig-hunters. McFarland knew a lepidopterist in Australia who was so skittish about letting colleagues find out which species he had and where he'd gotten them that he used a mirror to print his specimen labels backward—possibly with good reason: In a moment of indiscretion, another moth-man once told McFarland, "You know,

---

*Among McFarland's favorite authors is P. B. M. Allan, a British naturalist who wrote four books earlier in this century that make mothing sound almost sybaritic in its delights. Allan describes the art of painting sugar bait on tree trunks to attract moths in autumn: "And what kind of sugar must you use? . . . Brown Barbadoes, my lad, fresh from the docks, straight out of the ship's hold, moist and sticky . . . the sort of thing that causes every moth with a Latin name to exclaim, '*Odoratissimum mehercules saccharum!*'—'My hat, what glorious smelling sugar!' " Add to that "the particular brand of treacle which Mr. Fowler imports from the West Indies . . . one teaspoonful of *old* Jamaica rum, and stir well." Allan writes: "But the great advantage of autumn sugaring is that you can sugar at a convenient hour . . . and when you draw your chair up to the fire after dinner you can sip your port and go through the evening's doings at your ease, turning to your diary to see what moths you took this day last year and the year before that. . . . 'By Jove! It was this night five years ago that I took my first *rubiginea!*' "

Noel, the very best collecting is in other peoples' collections."
McFarland delights in the rivalries, but also stays close to his
collection.

"Oh, here's one of the wallbangers," McFarland said, as a
big, reddish moth came crashing through. "All saturniids live on
stored fat. They have vestigial mouthparts. These guys all have
real short tickets. They fly like crazy. They crash into branches.
They don't see real well and so in no time they're in tatters."
Then a geometrid moth he'd never seen before also turned up.
"You've got to be in a place for years and years to know what's
there," he said. He quoted Henry David Thoreau: "It takes a
man of genius to travel in his own country, in his native village,
to make any progress between his door and his gate." But it isn't
genius, he added. "It's staying home and doing a lot of work,
washing jars and changing caterpillar diapers."

McFarland's specialty is rearing moths from eggs, which
means figuring out what the caterpillars eat. "I give them a salad
of what's in leaf at the time they're out," he said. He arranges
his offerings in neat piles on a tray with Kleenex underneath,
hoping that he'll be able to find frass under one pile, meaning
that the caterpillars have found and eaten their natural food
plant. But it's rarely that simple. The caterpillars have often
evolved together with "some difficult, obscure little plant that's
out there, and if you keep trying you may find it." He once
spent four years desperately offering tray after tray before he
discovered the food plant for one species. Even then, the caterpillars often require constant nursing. McFarland has spent
much of his life fretting about whether the cage temperature is
too high or too low, and wondering if he should mist the caterpillars to keep them from drying out, or if one more squirt will
turn everything to mush. "I just like finding things out," he said.
"I don't give a damn whether it has any application or not. I just
like to solve the puzzle and then move on."

The puzzle often turns on the extraordinary ways one of those obscure little plants affects the moth's physical appearance and biochemistry. McFarland's book is full of photographs suitable for a child's game: What's wrong with this picture? Is that a casuarina twig? Or is it a caterpillar perfectly colored and striped and holding itself immobile all day long with precisely the curve of a casuarina twig? Among his backyard beasts in Arizona, McFarland in 1982 noticed one moth species so completely adapted to the oak that, in the spring, its caterpillars turn yellow and grow false pollen sacs, mimicking the tiny flowerlike catkins on the tree. But when the catkins fade, the very next generation of caterpillars grows up bumpy and gray, and these caterpillars spend their days studiously posing as oak twigs. The difference isn't merely skin deep: The flower-mimic caterpillars have small mouths and weaker musculature, for mumbling down the flower's soft pollen sacs. The twig caterpillars have big crusher jaws for eating leathery oak leaves. Before McFarland, scientists thought the caterpillars belonged to two different species. But both grow up to be the same emerald green looper moth.

These kinds of adaptations to a single plant are one reason any humble backyard can support such a menagerie of moths. The moths conquer by dividing and subdividing their habitat. For instance, many moths appear on the wing only during the brief period when their host plant is in flower. Even when they appear on the same night, different moths species often send out their mating calls at different set times. "The night looks uniform to us, but there are often windows of activity for different species," Dave Wagner says, adding that a good lepidopterist may be able to tell the date within two weeks and the time within an hour or two by which species are flying. The moths have at least one additional advantage for avoiding confusion among the host of moth species: They are acutely sensitive to the silent love songs sent out by members of their own species, usually composed of chemicals derived from the host plant.

Moth enthusiasts have known for centuries that moths can attract a mate over long distances; one venerable collecting technique was to tie a female moth to a tree and wait for the males to swoon in and court her. Scientists working with silkmoths at the Max Planck Institute in Munich began to explain this phenomenon in 1959, when they discovered, for the first time in any animal, one of the chemical signals now commonly known as pheromones.

A female silkmoth disperses just a billionth of a gram of her pheromone per hour and yet draws males from a mile or more away. The male's elaborate, comblike antennae are highly sophisticated olfactory sensors, sifting the air for minute traces of the pheromone. Each antenna has one thousand seven hundred hairs, and each hair has two thousand six hundred olfactory pores, all attuned to that one aphrodisiacal scent. The hairs concentrate the pheromone at a million times the level in the air itself, and the lovesick male comes running. (You can almost hear in the background the swelling chorus of the Rodgers and Hammerstein song: "Some enchanted evening, you may see a stranger, across a crowded room . . .") It makes a bloodhound snuffling up a hot scent trail look nose-dead. The scent is so potent that some unscrupulous spiders have learned to mimic moth pheromones as a way of luring males to their deaths.

Male moths also produce a scent. When the male gets close enough to the female, his body sprouts feather dusters and ticklers, loaded in some backyard species with the aromas of jasmine or cinnamon. These pheromones allow the female to determine if he is a suitable mate.

It's the "no cologne, no date" theory of moth mating, and Tom Eisner at Cornell University has described one of the most intriguing examples in Florida's ornate moth. Ornate moth caterpillars feed on a type of pea plant rich in an alkaloid toxin. This toxin, though powerful enough to kill a cow, is harmless to moths. They use it as a deterrent to birds and other predators.

But the really intriguing thing discovered by Eisner and his co-workers is that, when it comes time to mate, the female judges the male's desirability by how much of this poison he's accumulated. And the way she tells is from his cologne: The nastier he smells, the more she likes him.

But it gets even better: She'll mate with a half-dozen or so suitably toxic males. Each of these lucky guys puts about 11 percent of his body mass into the sperm package. It's the equivalent of a 180-pound man having a 20-pound ejaculate. But Eisner and his associate Craig LaMunyon discovered that only one of the males actually gets to fertilize her eggs. After copulation, the female determines which male really gets lucky, usually by selecting the largest sperm package. The other males get cuckolded. She uses the toxins they've given to her as nuptial gifts to anoint her eggs against predators.

It may be some consolation, if only to this writer, to know that male moths actually derive some benefit from carrying around all those toxins. It's not just energy spent in the cause of unrequited love. The plant poisons also protect the moths themselves: The scent of these moths is so distasteful that spiders actually cut them out of their webs and set them free. Plant toxins also discourage bats, which are the one significant disadvantage to flying at night.

Wagner, McFarland, and any other moth lover will sooner or later mutter dark and politically incorrect sentiments against bats, which linger around their black lights picking off moths. P. B. M. Allan, a British naturalist who wrote on the joys of mothing, once confessed to being so maddened by the sight of a pipistrelle seizing a prize moth that he got out his shotgun and blasted the bat into eternity. (Afterward, in a moment of horrified remorse, it occurred to him how threadbare the countryside

would be if the birds, bats, and other predators weren't out there busily keeping the moths and their caterpillars in check.)

But the moths are far from being defenseless against bats. They have evolved ears, enabling a moth to hear the bat's radar signals from one hundred feet off—as much as ten seconds before the bat even knows the moth is there. With visions of gaping mouths and needle-teeth dancing in its head, the moth flies away. If a bat gets close enough to race in for the kill, the moth can also hear the crescendo of ultrasonic shrieking. It may still save itself by folding its wings and making a last-ditch plunge.

It is a dogfight out there in the evening sky. Some moths even send out ultrasonic signals of their own, and researchers used to think the moths were jamming the bat's radar. But most now believe the moth is simply advertising the poisons it has picked up from its host plant. The bat gets the message, which is: "I taste rotten."

By the time I got back from visiting with Wagner and Mc-Farland to my own yard in Connecticut, it no longer seemed like the same place. I went to bed feeling like a boy who knows that his toy soldiers will begin to march the moment he closes his eyes for the night: Outside my window, the slender twigs on the ash tree would suddenly come to life and walk off on caterpillar legs. Strange poisons would ride the air in the causes of love and self-defense. Fragments of bark would take wing. Fallen leaves would rise back up to the treetops, and a million bird droppings would be reincarnated as lepidoptera.

By night, my yard now belonged, as it always has and probably always will, to the moths.

Slime

∽

"MONEY!" said the ad in *Pacific Fishing* magazine. Dollar bills fluttered through the air as tunas in top hats danced on their tail fins. "Do you want to come home every night and make $1,400 per day?"

The Pacific American Fish Company was looking for "a few good fishermen" in Astoria, Oregon, and Bellingham, Washington. I called up Peter at Pacific American and asked, "What's the catch?"

"Thirty cents a pound at the dock," he said, and a colleague reported that a single two-man boat out of Eureka, California, consistently brought in seven thousand pounds a night, from a trapline just three miles offshore. It sounded like a fisherman's dream. Better than a savings-and-loan scam! Safer than the stock market! Home every night! Ah, but the catch . . .

The catch, said Peter, was hagfish, a word to which one must always add, in a low, confidential tone, the explanatory phrase slime eels, or as fishermen in New England, where I live, like to say, slime hags, a term which just barely conveys the intended note of revulsion. Slime eels are bottom-dwelling deep-water scavengers, represented by about sixty species around the world, and before saying any more, I must issue a retraction. I once wrote ("Why Did God Make Flies?") that "it is hard to imagine an animal as disgusting or improbable as the housefly." It's clear now that I wasn't trying hard enough. Slime eels are more disgusting and improbable, and fishermen have known it for centuries.

Among other habits that have endeared them to seafarers, slime eels like to enter dead or dying bodies on the ocean bottom by way of mouth, gills, or anus, and gobble up everything except bones and skin, which remain intact. Fish immobilized in gill nets are particularly susceptible. In one study in the Gulf of Maine, slime eels gutted 3 percent to 5 percent of the catch. J. B. Heiser, a biologist at Cornell University's Shoals Marine Laboratory in Maine, describes what's left of the fish as "a bag of bones, literally . . . like it had been sucked dry by a high-powered vacuum cleaner."

Slime eels are often still inside the fish when the bloated gill net spills its contents onto the fisherman's deck, and Heiser, who has opened up several specimens, says the hags ensconced in their victim are typically well-fed and at ease, "smiling, slimy, usually snoring—gently." In one case, the record, a single cod contained 123 slime eels, in a pink, wriggling mass.

It is a disheartening sight for fishermen, touching some source of horror beyond mere economic loss. One fisheries expert has attributed this horror to the slime itself: "Being worthless . . . the hag is an unmitigated nuisance, and a particularly loathsome one owing to its habit of pouring out slime from its mucous sacs in quantity out of all proportion to its small size. One hag, it is said, can easily fill a two-gallon bucket, nor do we think this any exaggeration."

But a far graver problem with slime eels, it seems to me, is that they make the idea of burial at sea so much less appealing. Once having seen them, seafarers must suffer forever from foreboding that if they go down with the ship (or without it), slime eels will be waiting for them at the bottom. Martin Cruz Smith employed this idea to fine effect in his novel, *Polar Star,* in which a Russian murder victim returns from the bottom in a trawler net. As investigators examine a knife wound in the victim's gut, they notice a protruding "length of intestine, purplish-gray and slick," which gradually becomes recognizable as a slime eel: "The eel's head, an eyeless stump with fleshy horns and a puckered mouth, whipped from side to side against Zina Patiashvili's stomach; then the entire eel, as long as an arm, slid seemingly forever out of her, twisted in mid air" and landed at the examining physician's feet.

This scene is a figment of Smith's imagination. No one has ever found a human corpse being scavenged from within by slime eels. In truth, most researchers believe that even fish corpses are a relative rarity in the diet of slime eels, which are more likely to subsist on worms, shrimp, and other small sea-bottom creatures. But it is at least conceivable that hollowed-out shipwreck victims have at times drifted across the bottom like Michelangelo's self-portrait as a sack of skin on the wall of the Sistine Chapel, and the idea is enough to make one understand why Pacific American used dancing tunas, and not slime eels in top hats, for its appeal to fishermen. It may also explain the visceral reaction to slime eels in Smith's book, where the appearance of a slime eel on the fish-gutting line causes women to scatter "until the men had hammered it to death with shovels." It makes perfect sense that fishermen should hate slime eels. The real question is why anyone would want seven thousand pounds of these monsters delivered live every night.

The place to find out was Appledore Island, off the southern coast of Maine. Appledore used to be known as Hog Island before it gained status as a resort in the last century, and the polarized character of its identity has apparently stuck. The island is nowadays an object of medium-high cultural pilgrimage for gardeners and admirers of the American Impressionist Childe Hassam. Like Longfellow and other nineteenth-century luminaries, Hassam used to visit here at the home of a poet, Celia Thaxter, whose wallflowers and verbena are celebrated in her book, *An Island Garden,* recently republished. Modern visitors seeking to recapture that edifying era by and large avoid the unsightly World War II observation tower that now dominates the island, and this is probably just as well. Shoals Marine Laboratory uses the building for classes, and it contains a walk-in refrigerator that is the source of the island's current repute. J. B. Heiser refers to it as the "Hagfish Hilton" or, more often, "the Hagatorium." In the argot, Appledore itself sometimes becomes merely "Hagland."

Heiser, who is six feet, four inches tall, and his colleague Ric

Martini, who stands five-four, are the Click and Clack of the slime eel world. Heiser has grizzled blond curly hair and a beard, reddened skin, and slightly sea-bleached blue eyes; he has been director of Shoals Marine Laboratory for the past twelve years and talks in a stage voice audible over large portions of Appledore. Martini is trimmer and milder in manner, but also routinely delivers his shark-attack lecture while dressed as a great white shark. Shoals is a teaching facility, and they are educators more than researchers.

The two of them set out to catch their first slime eel ten years ago, when they realized that it was the only local species they could not actually show students in a marine vertebrate biology class. They soon became hooked on slime eels, as Heiser recently attested with the sort of forthright enthusiasm that scientists otherwise sternly repress: Slime eels, he said, "are the most disgusting, interesting organism I know of." What hooked Heiser and Martini, or at least gave them a pretext beyond sheer horrified fascination, were the slime eel's numerous biological anomalies. For example:

Slime eels possess seven rudimentary hearts and they breathe both through their gills and through a fine web of capillaries in the skin. But the skin doesn't bleed when cut, and the blood (when you can get it) is three times saltier than that of any vertebrate.

Slime eels exist in "mind-boggling" numbers on the sea floor, said Martini, "But it's not known when, where, or how they reproduce."

"We figure why is because it feels good," said Heiser. On the other hand, no one knows where the males are. A typical sample of 210 animals turned up only one.

"They don't have eyes," Martini interjected. "They don't have any eye musculature or the nerves that go with it. But you go in there and shine a flashlight on them and they immediately get up and dance." Slime eels, which live in gloomy depths down to six thousand feet or more, possess photoreceptor cells in their tails.

Biologists have typically used disparaging terms like "degeneration" and "regression" to sum up the slime eel's physiological endowment. Their "senses and the brain to which they pass their information tell us," one researcher has written, that the slime eel's life "is as empty as it is obscure." And he added, "Of course, it is of evolutionary interest that the survivors of perhaps the oldest stock of vertebrates should have retreated so far."

Slime eels undoubtedly deserve a certain amount of this characteristic human condescension. But one result has been that even scientists have long viewed slime eels inaccurately. Because slime eels lack conventional jaws, biologists have always classed them as second-rate lamprey eels, which are also jawless. (The mistake frequently extends to categorizing slime eels as parasites, rather than scavengers.) In truth, Heiser and Martini explained, there is little connection between slime eels and lampreys. Slime eels are not even vertebrates, but craniates. They were with us in the course of evolution long enough to develop a brain box, or cranium, at the end of a dorsal stiffening rod, called a notochord. But then they veered off on their own before the rest of us (or our slithery ancestors) got busy developing honest-to-God backbones. This suggests to Heiser and Martini that slime eels are not so much "degenerate" as merely primitive pre-vertebrates. For example, other scientists have argued that the slime eel's visual apparatus withered up and disappeared as it adapted to a burrowing, bottom-dwelling way of life. Heiser and Martini theorize that it may never have had visual apparatus in the first place—or even that it may have split off from the vertebrate evolutionary line before visual apparatus as we know it developed.

"I am convinced," said Heiser, "that the hagfish is the most primitive living relative of all vertebrate organisms. They're a door . . ."

". . . into the past," said Martini, in a mock-Gothic basso.

"AND they're disgusting," Heiser added, in case anyone had forgotten.

A fossilized impression of a slime eel, resembling the present-day animal down to the slime pores in its underbelly, exists from 220 million years ago. Some scientists look to slime eels of an even more-ancient date as the answer to what Stephen Jay Gould has called "the most famous and long-standing of all paleontological puzzles"—the animal origins of conodonts.

This is a famous puzzle, I should add, that only geologists have ever even heard of, but Gould's enthusiasm suggests its importance to them. Conodonts are small, toothlike structures widely distributed in rock of Cambrian and Triassic times, from 520 million to 200 million years ago. They are important because subtle variations over time enable geologists to date sedimentary layers to the relatively minute span of 500,000 years. Moreover, color changes in conodont deposits provide a record of temperature fluctuations in the surrounding rock. These traits have made conodonts vital to the search for gas and oil ("MONEY!" as the ad proclaimed). "There are major tomes written on these enigmatic pieces of dreck from the past," said Heiser.

"There are great phylogenetic lines," added Martini, "where this begat this, and this begat this. But they had no idea what animal these teeth belonged to." Some writers identified conodonts not as teeth but as plant parts, or as a copulatory structure in nematodes. Then, in 1982, paleontologists Euan Clarkson and Derek Briggs were studying a fossil shrimp bed in southern Scotland when they discovered a wormlike fossil that turned out to be a genuine conodont animal. Comparison of the teeth and other traits indicated that these animals resembled nothing quite so much as hagfish. Some paleontologists dispute this conclusion, pointing out that conodonts are chemically different from hagfish teeth. But general scientific opinion now holds that an animal much like the living slime eel, and possibly an ancestor, was swimming at the bottom of the Cambrian seas more than five hundred million years ago. Condescending humans are unlikely to last as long.

Studying the slime eel, said Heiser, is thus a bit "like planning the invasion of another planet. With no money."

"And not a whole lot of time," said Martini.

"And if we're right, that's exactly what we're doing. It's not a different planet. But it's this planet at a very different time."

In August 1989, Heiser and Martini descended four hundred feet in the Johnson Sea-Link II submarine to a sea-bottom projection known locally as the "Nipper," twelve miles off Appledore. On the first trip, they saw only a few indifferent slime eels, which was perhaps unsurprising; researchers who have studied them in captivity have characterized slime eels as slow, unresponsive, and generally inept—benthic morons lying about listlessly waiting for their chance to hitch a ride up to a factory ship and get hammered into oblivion by an outraged Russian fisherman.

On the second descent, the submarine brought along three pounds of stinking fish in a bucket. Two minutes, fifty-seven seconds after they touched bottom, Martini remarked, "Oh, my, yes." The pilot, uneasy about the idea that slime might shut down his thrusters, began furiously sucking up hagfish with a device like the pneumatic tube at a drive-in bank. Someone at the surface asked how many slime eels they were seeing. "Oh, I don' know," said the pilot, in the laconic, Chuck Yeager–drawl pilots always affect when they think they're about to buy the farm. "More 'n I can count." Say, four hundred. *Phwoop-phwoop-phwoop.* (Shoot, I would've been worried about my thrusters, too. But I was watching the whole thing after the fact, on videotape, amid the solid comforts of Appledore.)

In front of the submarine, slime eels continued to arrive, their pink heads up, sampling the water like schoolchildren enchanted by the scent of fresh-baked treats. The early arrivals crammed a small torpedo-shaped trap so full that some of them could get only their heads inside. A dozen tails wriggled out of one end like pink tentacles. Ropes and veils of slime drifted over everything. "If this is what three pounds of herring can do," said Martini, "just think what would happen if you were a gillnetter sitting here and you had a load of pollack."

The submarine finally took off, and when Martini looked down, he saw a carpet of slime eels fluttering below them. "It looks like Indiana Jones in the Temple of Doom," he said. ("They're supposed to swim languorously, and if they miss the bait they go limp," Martini remarked, as we watched the tape. "Obvious bullshit. These guys are steaming.")

At the surface, the submarine had to be deslimed with a firehose.

"Let's go meet some hagfish," Heiser declared, and, having prepared ourselves for the worst, we headed up to the Hagatorium, which was at first a disappointment. A researcher has written that one of the greatest obstacles to studying the behavior of slime eels "is the reluctance of this animal to behave at all under most laboratory conditions." Hence the tendency to underestimate their abilities. The Hagatorium was set at near-freezing temperature, which is the comfort zone for slime eels and also for Heiser, who worked in a sleeveless T-shirt and relied on what he called his "subcutaneous wet suit" for warmth. He and Martini had rigged a garden-hose circulation system to pump salt water through the fish boxes. But the slime eels were almost comatose in their immobility. They lay on the bottom of their tanks, stretched straight out, side by side, like two-foot-long hot dogs in the boiler of the corner hot dog cart. "Talk about exciting, dynamic animals!" said Martini. "People will want to take them home for pets." Heiser said he had persuaded slime eels to eat in captivity only once, when he ground up raw goldfish and served it with chopsticks, "like a sushi bar." Otherwise they mostly lay around with the listlessness often attributed to them. Martini reached in and plucked one out, and it quickly came to life, wriggling and gliding through his grip.

The East Coast species is known as *Myxine glutinosa,* a name which means "gluey slime," and it has ninety slime pores lined up like portholes on each flank. The pores blossomed into small white patches, and the slime festooned itself in ropes and formed a webbing between Martini's fingers. The pure volume of goo, Martini later explained, was a product of special slime cells, each containing a thread of protein balled up like a skein of yarn. As the protein unfolded, carbohydrate side chains branched out and gathered up surrounding water molecules to form a loose jelly. This molecular structure had a peculiar effect: It was easy enough to strip the slime off, but if you tried washing it off it merely became more voluminous.

Heiser explained that the slime is a defensive weapon. Any animal attempting to eat a slime eel (but what animal could be so depraved?) finds itself covered with a suffocating layer of the stuff and thus risks becoming a meal for its intended victim.

The slime can also suffocate the slime eel itself, and when Martini put it back in the water, it began to strip off the slime as assiduously as Lady Macbeth trying to wash the blood from her hands. It coiled itself into tight loops, then formed itself into a simple overhand knot, through which it drew its body. ("You'll see 'em tie a knot and then sneeze," said Heiser. "They backflush their gills and the slime comes out, *snoot*, from the nose." I'm sorry to report that I missed it.) Martini noted that slime eels also use knotting for purposes other than desliming. A slime eel will sometimes poke its head out of a trap (or a body) and tie a knot on the outside to apply force and pull its tail out. It can also attempt the opposite. Grabbing hold of something with its teeth, like a dog latched on to the seat of the mailman's pants, it knots its body and uses the knot to press against its prey, while tugging its head back through the loop. Slime eel teeth aren't built for ripping into large prey with ease (hence the propensity to use existing orifices). Having watched Heiser being bitten with impunity, I let a slime eel bite my bare hand, though I somehow could not bring myself to let it knot itself and apply full force. It was like being pinched hard with a folded piece of 50-grit sandpaper.

Having cleaned itself, Martini's slime eel swam around the tank with its outraged head just above the surface.

"There you see the adorable little piggy face," Martini commented.

In truth, what struck me at the moment was not its adorable pigginess, but the slime eel's unmistakable potential for human psychosexual trauma. Setting aside the question of burial at sea, was there any wonder that fishermen would have problems with a wriggling, snakelike animal that was phallic in color and shape and produced copious quantities of a substance distinctly resembling human semen? What if one also added that its eyeless, pink face—puckered, folded inward in a roughly star-shaped pattern—looked like a Pomeranian's anus? Or that it had yellow, comblike teeth arranged on a vertical axis, which flashed out without warning to grasp at food? Added to this, of course, was the whole necrophiliac business of entering dead bodies and devouring the innards. I neglected to ask Heiser and Martini if impressionable undergraduates ever came flying out of the Hagatorium yelling, "Mama, mama, mama, get me off this island and I'll never go near water again!" But I suspect that it is true.

It turns out that humans, at least, are depraved enough to eat slime eels, and this is where the commercial demand for them began. A slime eel fishery existed off Honshu in Japan through World War II, according to Susumu Kato, a researcher in the California office of the National Marine Fisheries Service, and it is still possible to buy live slime eels at retail markets in the South Korean port city of Pusan. In his report on a 1989 visit to Pusan, Kato noted his own reputation for "eating anything that swims or crawls on the ocean floor," and, as if this were sufficient explanation, added that he "just had to try hagfish." Friends took him to the most popular of several restaurants specializing in this dish. The bite-size pieces were charcoal-broiled with onions, chilis, and soybean paste, and garnished with chopped garlic. Kato commented that it "was not bad at all cooked in this fashion, though it helped to wash it down with good Korean wine."

Any food capable of being improved by Korean wine is probably not destined to find a market in this country. Of more compelling interest to Kato was the use of slime eel skins for leather. In this form, slime eels have been wriggling somewhere just under the American national consciousness since the early 1980s. Indeed, one could make a good case that, while greed, junk bonds, and leveraged buyouts all suggest solid possibilities, an apt and all-encompassing name for the 1980s might be "the slime eel decade."

According to Kato, South Korea perfected the process for

tanning slime eels in the late 1970s, yielding long, thin strips of leather, with a rich, supple feel and a slight bunching of flesh along the dorsal line. Factories began sewing them together to make belts, wallets, and purses. Sales of slime eel products to the United States boomed and the range of products rapidly expanded to include car phone covers, $250 briefcases, $1,000 golf bags, and other status symbols without which life in the 1980s would hardly have been worth living. Retailers generally marketed these products as "conger eel" or "Yuppie leather" (a name with the intriguing suggestion that it comes from recycled arbitrage specialists), to protect consumers from the awful knowledge that the thing cinched around their waist was once a slime eel. Importers I spoke to in Hawaii still insist that the leather they buy from Korea is "conger eel." But Kato gives the proper name as purple hagfish—that is, Pacific slime eel.

The slime eel business peaked, along with international financial markets, around 1986–1987, when there were six hundred boats and thirty-five leather processing plants in the business in Pusan alone. Under this intense fishing pressure, the average daily catch plunged from five tons per boat to about one ton, and boats began to go farther afield. Local hagfish populations also suffered because each boat routinely lost about two hundred traps a month. These traps, lacking any self-destruct mechanism, continued to "ghost fish" on the bottom. Along with the catch, the number of boats dropped, according to Kato, down to fewer than two hundred.

But demand for eel products remained strong enough that Korean companies fanned out around the world in search of new sources of hagfish, and in 1989 they imported 8.8 million pounds, much of it from the West Coast of the United States, where eighty boats entered the fishery. In 1990, Canada issued the first commercial licenses for hagfish trapping on the Atlantic coast.

. . .

One night I went out to the Nipper with Heiser and Martini, and off the rolling stern of the boat they lowered a weighted garbage can full of cold water and slime eels to the bottom. The slime eels were each marked with a small dot of indigo dye injected into a finfold, and the intent was to release them for recapture in future summers. The recaptures would help Heiser and Martini estimate the total slime eel population (a piece of information for which the whole world cries out in unison). When the empty garbage can was back on deck, we moved to another anchorage and sent down three traps baited with fish crud.

"We'll have hags in five minutes and within twenty we'll have all we want," said Heiser.

"That's amazing," said a passenger. "That must be the only kind of fishing where you get what you want."

"For a while." One aim of their research, Heiser said, was to find out something about slime eel reproduction and population before the local species suffers the kind of decimation that took place in the western Pacific. "Certainly, if you're going to make this an economically viable thing, you have to know enough to make it a sustained-yield fishery."

What he was suggesting was a highly rational, environmentally sound approach to a new resource. But somehow at that moment—perhaps it was my faith in humanity coming to the fore—the idea of a sustained-yield fishery in slime eels seemed unlikely. Like other status symbols of the 1980s, slime eel leather was in retreat. Factories were closing in Korea, and the importers I spoke with in Hawaii betrayed a skittishness about their future, which was troubled by false rumors and low-grade competition.

The rumors persisted from the late 1980s, when Yuppie leather was suspected of possessing mysterious electrical powers capable of erasing that decade's sacred text—the magnetic codes on credit cards. Television journalists phoned the nearest aquarium to demand if these could possibly be *electric* eels, and *USA Today* declared, "EELS ZAP US RIGHT IN OUR POCKET-BOOKS." It later developed that a magnetic clip on eelskin purses, rather than the eelskin itself, was causing the problem. A more serious setback occurred, judging from the comments of importers, when Korean tanneries began to accept slime eels from North America, Japan, and other fisheries. These new varieties of Yuppie leather tended to rub right through in a month, and the dyes did not hold. Shoddy goods were damaging the reputation even of the Korean original.

At the back of the boat, Heiser and Martini began winching up their traps, and all too soon, three seething nests of slime eels arrived on deck. I could not help thinking that the real problem for Yuppie leather was that the 1990s were shaping up as a colder, harder decade, wherein people must perforce look past the patina on the wallet and confront the empty billfold and overextended credit cards within. And when they confronted the reality of slime eels . . .

The way I imagined it, *Polar Star* would become a hit movie and the big slime eel scene would have an impact in inverse proportion to the Reese's Pieces scene in *E.T.* The Yuppie leather empires of the world would come crashing down, fishermen (who never really expected to make $1,400 a day anyway) would turn to some less troublesome catch, and slime eels would be left to get on with their dreary business on the ocean bottoms of the world, where, with luck, we would never see them again.

Bibliography

Introduction: The Joy of Formication

Barnes, R. D., and E. E. Rupert. *Invertebrate Zoology*, 6th ed. Lithonia, Ga.: SCP: Third World Literature, 1993.

Berenbaum, M. R. *Bugs in the System: Insects and Their Impact on Human Affairs*. New York: Addison-Wesley, 1995.

Fabre, J. H. *The Insect World of J. Henri Fabre*. New York: Dodd, Mead, 1949; reprint 1981.

Milne, L., and M. Milne. *National Audubon Society Field Guide to North American Insects and Spiders*. New York: Knopf, 1980.

Preston-Mafham, R., and K. Preston-Mafham. *The Encyclopedia of Land Invertebrate Behaviour*. Cambridge, Mass.: MIT Press, 1993.

Taylor, R. L., and B. J. Carter. *Entertaining with Insects: or, The Original Guide to Insect Cookery*. Santa Barbara, Calif.: Woodbridge Press, 1976.

Wilson, E. O. *The Diversity of Life*. New York: Norton, 1992.

———. *The Insect Societies*. Cambridge, Mass.: Harvard University Press, 1971.

Bibliography

Why Did God Make Flies?

Dethier, V. G. *The Hungry Fly.* Cambridge, Mass.: Harvard University Press, 1976.

Greenberg, B. *Flies and Disease,* Vol. 2. Princeton, N.J.: Princeton University Press, 1973.

Nachtigall, W. *Insects in Flight.* New York: McGraw-Hill, 1992.

Oldroyd, H. *The Natural History of Flies.* New York: Norton, 1964.

West, L. S. *The Housefly: Its Natural History, Medical Importance, and Control.* Ithaca, N.Y.: Comstock Publishing, 1951.

Little Suckers

Drapeau, P., W. Kristan, and M. Shankland, eds. "Neurobiology and Development of the Leech." *J. of Neurobiology* 27 (1995): 267–445.

Klemm, D. J. *Leeches (Annelida: Hirudinea) of North America.* Cincinnati, Ohio: U.S. Environmental Protection Agency, 1982.

———, ed. *A Guide to the Freshwater Annelida (Polychaeta, Naidid and Tubificid Oligochaeta, and Hirudinea) of North America.* Dubuque, Iowa: Kendall/Hunt, 1986.

Kristan, W. B., Jr. "The Neurobiology of Swimming in the Leech." *Trends in Neuroscience* 6 (1983): 84–88.

Nicholls, J. G., and D. Van Essen. "The Nervous System of the Leech." *Scientific American* 230 (1974): 38–48.

Sawyer, R. T. *Leech Biology and Behaviour.* Oxford: Oxford University Press, 1986.

Stent, G. S., and D. A. Weisblat. "The Development of a Simple Nervous System." *Scientific American* 246 (1982): 136–146.

Empire of the Fire Ant

Daniel, P. "A Rogue Bureaucracy: The USDA Fire Ant Campaign of the Late 1950s." *Agricultural History* 64[2] (1990): 99–114.

Holldöbler, B., and E. O. Wilson. *The Ants.* Cambridge, Mass.: Belknap Press, 1990.

Revkin, A. C. "March of the Fire Ant." *Discover* 10 (1989): 70–76.

Tschinkel, W. R. "The Ecological Nature of the Fire Ant: Some Aspects of Colony Function and Some Unanswered Questions." In *Fire Ants and Leaf-cutting Ants: Biology and Management.* Boulder: West-view Press, 1986.

————. "Fire Ant Queen Longevity and Age: Estimation by Sperm Depletion." *Annals of the Entomological Society of America* 80 [2] (1987): 263–266.

Tschinkel, W. R., and S. D. Porter. "Efficiency of Sperm Use in Queens of the Fire Ant, *Solenopsis invicta* (Hymenoptera: Formicidae)." *Annals of the Entomological Society of America* 81 [5] (1988): 777–781.

Vinson, S. B., and L. Greenberg. "The Biology, Physiology, and Ecology of Imported Fire Ants," In *Economic Impact and Control of Social Insects.* New York: Praeger, 1986.

Yoffe, E. "Ants from Hell." *Texas Monthly* August (1988): 80ff.

The Big Calamari

Boyle, P. R. "Neural Control of Cephalopod Behavior," In *The Mollusca Vol. 9: Neurobiology and Behavior, Part 2.* New York: Academic Press, 1986.

Corning, W. C., J. A. Dyal, and A. O. D. Willows, eds. *Invertebrate Learning Vol. 3: Cephalopods and Echinoderms.* New York: Plenum Press, 1973.

Herring, P. J., P. N. Dilly, and C. Cope. "The Bioluminescent Organs of the Deep-Sea Cephalopod *Vampyrotheuthis infernalis* (Cephalopoda: Vampyromorpha)." *J. Zool.* 233 (1994): 45–55.

Roper, C. F. E., and K. J. Boss. "The Giant Squid." *Scientific American* 246 [4] (1982): 96–105.

Roper, C. F. E., M. J. Sweeney, and C. E. Nomen. "Cephalopods of the World," In *FAO Fisheries Synopsis No. 125, Vol. 3.* New York: United Nations Development Programme, 1984.

Wells, M. J. "Oxygen Extraction and Jet Propulsion in Cephalopods." *Can. J. Zool.* 68 (1990): 815–824.

Lions of the Pond

Carpenter, V. *Dragonflies and Damselflies of Cape Cod.* Provincetown, Mass.: Cape Cod Museum of Natural History, 1991.

Dunkle, S. W. *Damselflies of the Florida Peninsula, Bermuda, and the Bahamas.* Gainesville, Fla.: Scientific Publishers, 1990.

———. *Dragonflies of the Florida Peninsula, Bermuda, and the Bahamas.* Gainesville, Fla.: Scientific Publishers, 1989.

Montgomery, B. E. "Why Snakefeeder? Why Dragonfly? Some Random Observations on Etymological Entomology." *Proceedings of the Indiana Academy of Science* 82 (1973): 235–241.

Preston-Mafham, R., and K. Preston-Mafham. *The Encyclopedia of Land Invertebrate Behaviour.* Cambridge, Mass.: MIT Press, 1993.

Waage, J. K. "Adaptive Significance of Post-copulatory Guarding of Mates and Nonmates by Male *Calopteryx maculata* (Odonata)." *Behav. Ecol. Sociobiol.* 6 (1979): 147–154.

———. "Dual Function of the Damselfly Penis: Sperm Removal and Sperm Transfer." *Science* 203 (1979): 916–918.

Spider Love

Bristowe, W. S. *The World of Spiders.* London: Collins, 1958; revised 1971.

Fawcett, P. H. *Lost Trails, Lost Cities* (published in England under the title *Exploration Fawcett*). New York: Funk & Wagnalls, 1953.

Foelix, R. F. *Biology of Spiders.* Cambridge, Mass.: Harvard University Press, 1982.

Gertsch, W. J. *American Spiders.* New York: Van Nostrand Reinhold, 1949.

Martins, M. "Why Do Snakes Sleep on the Vegetation in Central Amazonia?" *Herpetological Review* 24 [3] (1993): 83–84.

Tu, A. T., ed. "Insect Poisons, Allergens, and Other Invertebrate Venoms." In *Handbook of Natural Toxins,* Vol. 2. New York: Dekker, 1984.

Leapers

Busvine, J. R. *Insects, Hygiene, and History.* London: Athlone Press, 1976.

Lehane, B. *The Compleat Flea.* New York: Viking Press, 1969.

Marshall, A. G. *The Ecology of Ectoparasitic Insects.* London: Academic Press, 1981.

Rothschild, M. "Fleas." *Scientific American* 213 [6] (1965): 44–53.

Rothschild, M., and T. Clay. *Fleas, Flukes & Cuckoos.* London: Arrow, 1957.

Rothschild, M., et al. "The Flying Leap of the Flea." *Scientific American* 229 [5] (1973): 92–100.

Traub, R., and H. Starcke, eds. *Fleas*. Rotterdam: A. A. Balkema, 1980.

Beetlemania

Cowan, F. *Curious Facts in the History of Insects*. Philadelphia: Lippincott, 1865.

Erwin, T. L. "Tropical Forest Canopies: The Last Biotic Frontier." *Bulletin of the Entomological Society of America* Spring (1983): 14–19.

———. "Tropical Forests: Their Richness in *Coleoptera* and Other Arthropod Species." *Coleopterists' Bulletin* 36 [1] (1982): 74–75.

Forsyth, A., and K. Miyata. *Tropical Nature*. New York: Scribner's, 1984.

Klausnitzer, B. *Beetles*. New York: Exeter, 1981.

Grunting for Wigglers

Appelhof, M. *Worms Eat My Garbage*. Kalamazoo, Mich.: Flower Press, 1982.

Appelhof, M., M. F. Fenton, and B. L. Harris. *Worms Eat Our Garbage: Classroom Activities for a Better Environment*. Kalamazoo, Mich.: Flower Press, 1993.

Darwin, C. *The Formation of Vegetable Mould, Through the Action of Worms, with Observations on Their Habits*. New York: MAS Press, 1972.

Edwards, C. A., and J. R. Lofty. *Biology of Earthworms*, 2nd ed. New York: Wiley, 1977.

Laverack, M. S. *The Physiology of Earthworms*. Oxford: Pergamon Press, 1963.

Macdonald, D. W. *Running with the Fox*. London: Unwin Hyman, 1987.

Satchell, J. E., ed. *Earthworm Ecology: From Darwin to Vermiculture*. London: Chapman & Hall, 1983.

A Small Point of Interest

Bates, M. *The Natural History of Mosquitoes*. New York: Macmillan, 1949.

Booth, W. "AIDS and Insects." *Science* 237 (1987): 355–356.

Busvine, J. R. *Insects, Hygiene, and History.* London: Athlone Press, 1976.

Fisher, M., and D. E. Fisher. "The Attack of the Killer Mosquitoes." *Los Angeles Times Magazine* Sept. 15 (1991): 31–35.

Gillet, J. D. *The Mosquito: Its Life, Activities, and Impact on Human Affairs.* Garden City, N.Y.: Doubleday, 1972.

Harrison, G. A. *Mosquitoes, Malaria, and Man: A History of the Hostilities Since 1880.* New York: Dutton, 1978.

Powell, J. H. *Bring Out Your Dead: The Great Plague of Yellow Fever in Philadelphia in 1793.* Philadelphia: University of Pennsylvania Press, 1949.

Ghosts on Wings

Allan, P. B. M. *Leaves from a Moth Hunter's Notebook.* Oxford: E. W. Classey, 1949; reprint 1975.

———. *A Moth Hunter's Gossip.* London: Watkins & Doncaster, 1947.

———. *Moths and Memories.* London: Watkins & Doncaster, 1948.

———. *Talking of Moths.* Oxford: E. W. Classey, 1943; reprint 1975.

Covell, C. V., Jr. *A Field Guide to the Moths of Eastern North America.* Boston: Houghton Mifflin, 1984.

Holland, W. J. *The Moth Book.* New York: Doubleday, 1903; reprinted with updating by A. E. Brower. New York: Dover, 1968.

McFarland, N. *Portraits of South Australian Geometrid Moths.* Lawrence, Kans.: Allen Press, 1988.

Quammen, D. *Natural Acts.* New York: Nick Lyons Books, 1985.

Stowe, M. K., J. H. Tumlinson, and R. R. Heath. "Chemical Mimicry: Bolas Spiders Emit Components of Moth Prey Species Sex Pheromones." *Science* 236 (1987): 964–967.

Waage, J. K., and G. G. Montgomery. "*Cryptoses choloepi:* A Coprophagous Moth That Lives on a Sloth." *Science* 193 (1976): 157–158.

Slime

Aldridge, R. J. "A Soft Body of Evidence." *Natural History* May (1989): 6+.

Fange, A. B., and R. Fange, ed. *The Biology of Myxine.* Oslo, Norway: Universitetsforlaget, 1960.

Pough, F. H., J. B. Heiser, and W. N. McFarland. *Vertebrate Life,* 4th ed. Upper Saddle River, N.J.: Prentice-Hall, 1996.

Smith, Martin Cruz. *Polar Star.* New York: Random House, 1989.

Worthington, J. "Contribution to Our Knowledge of Myxinoids." *Amer. Nat.* 39 (1905): 625–663.

INDEX

Page numbers in *italics* refer to illustrations.